MY EUROPEAN FAMILY

Also available in the Bloomsbury Sigma series:

Sex on Earth by Jules Howard
p53: The Gene that Cracked the Cancer Code by Sue Armstrong
Atoms Under the Floorboards by Chris Woodford
Spirals in Time by Helen Scales
Chilled by Tom Jackson
A is for Arsenic by Kathryn Harkup
Breaking the Chains of Gravity by Amy Shira Teitel
Suspicious Minds by Rob Brotherton
Herding Hemingway's Cats by Kat Arney
Electronic Dreams by Tom Lean
Sorting the Beef from the Bull by Richard Evershed
and Nicola Temple
Death on Earth by Jules Howard
The Tyrannosaur Chronicles by David Hone
Soccermatics by David Sumpter
Big Data by Timandra Harkness
Goldilocks and the Water Bears by Louisa Preston
Science and the City by Laurie Winkless
Bring Back the King by Helen Pilcher
Furry Logic by Matin Durrani and Liz Kalaugher
Built on Bones by Brenna Hassett

MY EUROPEAN FAMILY

THE FIRST 54,000 YEARS

Karin Bojs

BLOOMSBURY
sigma

Bloomsbury Sigma
An imprint of Bloomsbury Publishing Plc

50 Bedford Square
London
WC1B 3DP
UK

1385 Broadway
New York
NY 10018
USA

www.bloomsbury.com

BLOOMSBURY and the Diana logo are trademarks of Bloomsbury Publishing Plc

First published in Sweden in 2015 by Albert Bonniers Forlag, as *Min Europeiska Familj* by
Karin Bojs. This edition published in the United Kingdom in 2017 by Bloomsbury Publishing.
Published by agreement with the Kontext Agency.

Copyright © Karin Bojs, 2017
English language translation © Fiona Graham, 2017

Photo credits (t = top, b = bottom, l = left, r = right, c = centre)
Colour section: P. 1: Heritage Images/Gettyimages (tl); AFP/Stringer/Gettyimages (cr, b).
P. 2: Jens Schlueter/Staff/Gettyimages (t); Zhengan/Creative Commons Attribution-Share
Alike 4.0 International (cl); Print collector/Getty images (bl). P. 3: AFP/Stringer/Gettyimages
(tr); Sovfoto/Getty images (b). P. 4: Heritage Images/Gettyimages (tl); Klaus D. Peter/Creative
Commons Attribution-Share Alike 4.0 International (cl); Crochet.david/Creative Commons
Attribution-Share Alike 4.0 International (b). P. 5: *Science* 2014, Niels Bohr Institute/Stefan
Rothmaier (t); Haak et. al., *Nature* 2015, Allentoft et. al., *Nature* 2015/Stefan Rothmaier (b).
P. 6–7: Łukasz Lubicz Łapiński/Stefan Rothmaier. P. 8: Tomas Larsson (tr);
Gunnar Källén (cr); Karin Bojs (b).

Karin Bojs has asserted her right under the Copyright, Designs and
Patents Act, 1988, to be identified as Author of this work.

British Library Cataloguing-in-Publication data
A catalogue record for this book is available from the British Library.

Library of Congress Cataloguing-in-Publication data has been applied for.

ISBN (hardback) 978-1-4729-4147-3
ISBN (trade paperback) 978-1-4729-4146-6
ISBN (ebook) 978-1-4729-4149-7

2 4 6 8 10 9 7 5 3 1

Diagrams by Stefan Rothmaier

Typeset by Deanta Global Publishing Services, Chennai, India
Printed and bound in Great Britain by CPI Group (UK) Ltd, Croydon CR0 4YY

Bloomsbury Sigma, Book Twenty-one

MIX
Paper from
responsible sources
FSC® C020471

To find out more about our authors and books visit www.bloomsbury.com.
Here you will find extracts, author interviews, details of forthcoming
events and the option to sign up for our newsletters.

To Anita and Göran,
from whom I inherited all my genes.

Contents

Introduction: The Funeral   9

PART 1: THE HUNTERS

Chapter 1: The Troll Child: 54,000 Years Ago   15

Chapter 2: Neanderthals in Leipzig   18

Chapter 3: The Flute Players   38

Chapter 4: First on the Scene in Europe   47

Chapter 5: Mammoths in Brno   57

Chapter 6: Cro-Magnon   62

Chapter 7: The First Dog   78

Chapter 8: Doggerland   86

Chapter 9: The Ice Age Ends   94

Chapter 10: Dark Skin, Blue Eyes   110

Chapter 11: Climate and Forests   118

Chapter 12: Am I a Sami?   124

Chapter 13: Pottery Makes its Appearance   134

Chapter 14: The Farmers Arrive   139

PART 2: THE FARMERS

Chapter 15: Syria   145

Chapter 16: The Boat to Cyprus   147

Chapter 17: The First Beer   158

Chapter 18: The Farmers' Westward Voyages   173

Chapter 19: The Homes Built on the
Graves of the Dead 181

Chapter 20: Clashes in Pilsen and Mainz 189

Chapter 21: Sowing and Sunrise 196

Chapter 22: Farmers Arrive in Skåne 203

Chapter 23: Ötzi the Iceman 208

Chapter 24: The Falbygden Area 214

Chapter 25: Hunters' and Farmers' Genes 230

PART 3: THE INDO-EUROPEANS

Chapter 26: The First Stallion 239

Chapter 27: DNA Sequences Provide Links
with the East 254

Chapter 28: Battleaxes 267

Chapter 29: Bell Beakers, Celts and Stonehenge 275

Chapter 30: The Nebra Sky Disc in Halle 281

Chapter 31: The Rock Engravers 287

Chapter 32: Iron and the Plague 301

Chapter 33: Am I a Viking? 307

Chapter 34: The Mothers 326

Chapter 35: The Legacy of Hitler and Stalin 340

The Tree and the Spring 349

Questions and Answers about DNA 352

References, Further Reading and Travel Tips 358

Acknowledgements 394

Index 395

The Funeral

My mother, Anita Bojs, died while I was working on this book. Many friends and acquaintances came to the funeral, considerably more than I had dared to hope for. But there were only a few family members. All of us fitted into one pew at the front: my brother and me, our partners, and three smartly dressed grandchildren.

It was early summer, and the mauve lilac was in bloom in the park next to Gothenburg's Vasa Church. Together we sang the hymn 'Den blomstertid nu kommer' ('The blossom time is coming'), followed by 'Härlig är jorden' ('Fair is Creation'). I had chosen the latter because it includes two lines that I find particularly comforting: 'Ages come and ages pass / Generation follows upon generation.'

When the time came for the memorial speeches, I addressed myself particularly to Anita's grandchildren. I wanted them to feel proud of their grandmother and her origins, despite the circumstances.

The grandmother they had known was elderly and scarred by life. Her promising career had come to an end when she was still in her fifties. By then she already had a long history of serious problems including illness, divorce, conflicts and addiction.

That was why I told her grandchildren and the other people at the funeral about the first half of my mother's life. I told them about the student who scored top marks at school and landed a place to study medicine at the Karolinska Institute – although she was a girl and came from a family of modest means. I told them about her childhood home above the little school in the industrial village where my grandmother Berta taught. That home had been poor in financial terms, but it was all the richer in companionship, music, art, literature and the thirst for knowledge.

I quoted from a diary I had found among my mother's things. 'Top Secret' was written on the cover in a childish hand. In the diary she had written about her summer holidays in the province of Värmland, at the home of her grandmother, my great-grandmother Karolina Turesson. It was just 40 kilometres (25 miles) away from the border with Norway, where the Second World War was raging. Yet the description of the cousins playing together on the jetty at glittering Lake Värmeln was so bright and full of beauty in retrospect.

I never met my maternal grandmother or great-grand-mother. With all the problems there were in my troubled family, I seldom met any relatives at all. That may be why I have spent so much time wondering who these relatives were and where they came from. I have wanted to know more since I was a 10-year-old.

On my father's side, at least I met my grandmother Hilda and my grandfather Eric a few times. My visits to them are among my happiest memories. Their home in Kalmar smelt so good. There were paintings everywhere, and my grand-father had decorated doors, walls and furniture with pictures of his own. Both grandparents were fond of telling stories from their childhood, but they rarely said anything about family further back in the past.

Now, as an adult, expert genealogists have helped me to trace Berta's, Hilda's and Eric's ancestry over many generations. When I was young, I thought it was rather nerdy to research your family history. But as an adult I now have far more respect for the activity. I understand how a fascination with one's forebears is an important component of many of the world's cultures. In many of the ethnic groups that lack a written tradition, people can rattle off their ancestry for at least 10 generations – about as far back as successful family history researchers in Sweden. The Bible sets out lengthy genealogies, the oldest of which were written down over 2,000 years ago, having being passed on by oral tradition long before that.

Methods for tracing one's ancestry have moved on since biblical times. In fact, I would say the current unprecedented developments are nothing less than quantum leaps. About 50 years ago, a few pioneering researchers began to compare blood groups and individual genetic markers in order to be able to identify family connections and historical migrations. At that time, the DNA molecule had only been known for a few years, and the relevant knowledge was shared by a very limited group of researchers. It was not until 1995 that it became possible to examine all the DNA of a tiny bacterium. Since then, progress has been astonishingly rapid. In fact, biotechnology is undergoing even more dramatic developments than information technology, although advances in computing, telephones and the internet are more visible to the general public. IT people speak of 'Moore's law', according to which computers double their capacity every two years. The capacity to map a DNA sequence is advancing much faster than that.

A few years ago, it became possible to analyse the whole of a person's genetic make-up in a matter of hours. Researchers can even analyse DNA from people who have been dead for tens of thousands – and, in one or two cases, hundreds of thousands – of years. An analysis that would have cost tens of millions of pounds a decade or two ago is now obtainable for less than £50. Thanks to lower prices, this technology has also begun to spread beyond the world of professional research. Even private individuals researching their family tree have started to use DNA as a tool. Small variations in the DNA sequence make it possible to identify unknown first, second and third cousins, and even family members who lived a very long time ago – during the last Ice Age and even further back.

I have followed developments in DNA technology as a science journalist over the last 18 years, for most of which I have been science editor at *Dagens Nyheter*, one of Sweden's leading daily newspapers. I have witnessed a revolution in the technology used in criminology and medical and biological research, and I have seen how DNA technology is now beginning to contribute to archaeology and history as well.

This book is an attempt to gather together these different threads: professional researchers' latest findings about the early history of Europe, and my own particular family history. My research has involved travelling in 10 countries, reading several hundred scientific studies and interviewing some 70 researchers.

I am now beginning to be able to discern the threads that link my maternal great-grandmother Karolina, my paternal grandmother Hilda and my paternal grandfather Eric, and events that took place a long way back in history. Most of these threads are shared with a large part of the population of Europe.

Let us begin with something that happened near the Sea of Galilee, about 54,000 years ago.

PART ONE
THE HUNTERS

Annika gave birth to Märta, who gave birth to Karin.

Karin was the mother of Annika, who herself had a daughter named Karin.

Karin's daughter was Kajsa, and Kajsa gave birth to Karolina.

Karolina gave birth to Berta, who gave birth to Anita.

Anita gave birth to Karin.

CHAPTER ONE

The Troll Child:
54,000 Years Ago

The woman who was to become my ancestor strode swiftly down the mountainside. She was in a hurry to reach the lake in the valley, which we today know as the Sea of Galilee. At that time the lake covered a far larger area. Contemporary geologists have retrospectively named it Lake Lisan ('tongue') on account of its elongated shape. What name the woman and her group gave to that great expanse of water is something we shall never know.

She was young and slender with curly black hair, and her skin was a dark brown. She wore nothing but a twisted cord about her hips, from which dangled strings of shells dyed red and green. They swung in time with her strides. Between her naked breasts hung a strip of leather bearing an amulet – the small figure of a bird, made of gazelle horn.

On the mountain the woman had met a man whose seed was now in her belly.

The man was about the same height as the woman, but much heavier and more powerfully built. Their encounter may have been violent, in which case she could hardly have been in a position to defend herself effectively. His face was not like that of anyone the woman had previously encountered; his nose was much larger and broader, and the whole of his face seemed to jut forward. His eyes were brown, just like hers, but his skin was lighter and his hair straight. Above all, he had a characteristic scent, which she had not noticed until he came close. It was an acrid, alien smell.

Although the man and the woman were so unalike, a child began to grow within her. By the time she was due to give birth, winter had come. Though the small band of people she belonged to were still on the shores of Lake Lisan, they had moved a few dozen kilometres southwards, towards their old homeland. They put up some simple windbreaks along a cliff face, hoping the creatures they had found living in the new country would leave them in peace. The Others – or the trolls, as they sometimes called them.

A chill rain fell. Now the ground was bare of the red tulips of spring time: nothing but a few dry thistles remained along the shoreline. The birth was difficult, but both the woman and her baby survived. It was a boy – an unusually big, robust child. My ancestor wrapped him in a gazelle hide and laid him carefully on a bed of dried grass.

Three days later the shaman performed a ritual, dancing frenetically until she made contact with the gods, while the other members of the group sat around the fire chanting. On returning from the world of the gods, the shaman had much to say about the child's future. She foretold that he would have many descendants. They would be dispersed in all directions, even to the ends of the earth. 'He shall be called the son of the gods, and the gods will give you strength enough to raise him,' the shaman told the young mother.

It was most unusual for her to lavish such extraordinary words on newborn babies, but the shaman was a wise woman. She could see that the woman who was to become my ancestor would need to have a special vocation conferred on her if the child was to survive. The group could not afford to lose any more children, or there would be an end to their presence in the new country.

Fortunately, the boy turned out robust and strong. He had a healthy appetite, first for his mother's milk, then for shreds of meat and plants. He drank out of brooks, and never even suffered from an upset stomach. But he was not quite like other children in appearance. His skin was lighter. His chin was smaller, and it receded towards his throat. His eyebrow ridges were more pronounced. When his hair started to grow longer, it hung straight. And when his mother was out of earshot, the other members of the group often called him the 'Troll Child'. It was said with affection, but they had their suspicions about his origins.

After a while they plucked up the courage to return to the mountains in the north, even though The Others were there. The region we now call Galilee provided such favourable living conditions, abounding in gazelles, aurochs and other game. They found a fine, spacious limestone cave that provided shelter in winter. Only rarely did they glimpse one of The Others, and that was nearly always at a long distance. On one or two occasions they got within hearing range. The Others had a strange way of speaking, and it was impossible to make head or tail of their language. Their garments were plain and unadorned.

The Troll Child was not at all quick in starting to talk, nor did he enjoy stories as ordinary children did. Yet he was at least as skilful with flint, wood and animal hides as anyone else. He just needed to be shown what to do.

During this time the winters became unusually cold, wet and harsh. Many small children fell sick, and several died. But the Troll Child thrived, growing stronger and stronger. The group took good care of him. They gave him extra attention, as was their wont with anyone who was a little different. Moreover – though they never actually voiced this fear – they were anxious that someone might come and take him away from them.

A few years later the woman died, without ever having told anyone about the man she had met that spring day when the tulips bloomed red on the mountain slopes. But the words of the shaman came true. The Troll Child did have a large number of descendants, and they were dispersed in all directions, even to the ends of the earth.

CHAPTER TWO

Neanderthals in Leipzig

M ost of the human beings alive today have one or two 'troll children' among our forebears. A small proportion of our genetic material comes from 'The Others' – the people we now know as Neanderthals. This share is slightly less than 2 per cent for people of European descent like myself – the same proportion as if my grandmother's grandmother had had a grandfather who was a Neanderthal. It is as though a 'troll child', fathered by a Neanderthal, were the father of my grandmother's grandmother.

Naturally that isn't the case. The father of my grandmother's grandmother lived in the nineteenth century. The inter-breeding of Neanderthals and modern humans, however, occurred much earlier – about 54,000 years ago.

Yet the traits inherited from the Neanderthals have remained with us for thousands of years because the population was so tiny at that time. This meant that a few instances of interbreeding with Neanderthals had a major impact. Moreover, some of the characteristics associated with the Neanderthals were useful, improving people's chances of surviving and having children of their own.

The scene of our sexual interaction with Neanderthals is likely to have been the Middle East, a corridor through which all humans passed on their way from Africa to other parts of the world. It may well have been Galilee, which archaeological finds show to have been inhabited concurrently by modern humans and Neanderthals. Alternatively, it may have been a little further to the north, possibly in what is now Lebanon.

The Neanderthals were the first to arrive in the region, their ancestors having left Africa several hundred thousand years before us. Traces of Neanderthals can be found from Spain in the west to Siberia in the east. One of the first finds was made in the mid-nineteenth century, in Germany's Neander Valley, hence the name 'Neanderthals'.

From the discovery in the Neander Valley right up to the last few decades, most researchers believed that today's Europeans were the Neanderthals' grandchildren, figuratively speaking. They thought we had developed here, isolated from other peoples for a long period of time, and had thus developed a typically European appearance, with light skin and straight hair. People in Asia and Africa were thought to have had separate developmental histories; Asians, for example, were believed to be descended from other prehistoric peoples such as Peking Man and Java Man. These ideas, known as the multi-regional hypothesis, are largely mistaken. Yet they do contain a few small grains of truth, as revealed by new DNA technology.

***

Through his studies of ancient DNA, Swedish-born Svante Pääbo has done more than anyone else to shed light on the early history of humankind. Today he is one of the world's most renowned scientists and the director of the department of evolutionary genetics at the Max Planck Institute for Evolutionary Anthropology, of which he is a co-founder, in the German city of Leipzig.

I visited the Institute twice while working on this book. It is housed in a specially designed, spacious building, with light flooding in through glass walls. A pool encircled by green plants gleams in a central atrium. At the entrance there is a climbing wall four storeys high, built to Pääbo's specifications. This is used by young researchers training for African fieldwork on treetop-dwelling monkeys. Next to the climbing wall stands a grand piano which is used for choir practice. These features go to show some of the Institute's unique character. Researchers here work in disciplines as diverse as psychology, palaeontology and linguistics, seeking to develop an understanding of how we came to be the people we are today. But it is molecular biology and DNA research that lie at the very heart of the Institute's work.

The special laboratory for extracting ancient DNA is situated in the basement, to avoid any unwanted contamination. It is here that Pääbo and his young colleagues strive to further refine their technology. Although they now have many competitors in other countries, the Leipzig group is still a world leader. During the week when I visited the Institute for the second time, they published a study on DNA from an archaic human found in Spain, some 400,000 years old – from a time before even the Neanderthals had emerged.

Pääbo's work is cutting-edge research that drives progress in technology and knowledge. But the fact remains that it was the Neanderthals that made him famous, and it is his increasingly high-resolution analyses of Neanderthal DNA that have made him known to the general public.

When I entered Pääbo's office, the first thing to meet my eyes was a Neanderthal skeleton, squeezed in between the desk and the sofa. It was a composite, made up of copies of various excavated bones. In contrast to the short, powerfully built Neanderthal, Pääbo was tall and lanky, with a long, narrow face.

Newspapers worldwide have dedicated metres of column space to his research and his unusual family background as the secret, illegitimate son of Sune Bergström, the Nobel Prize-winning vice-chancellor (or president, in US terminology) of Stockholm's Karolinska Institute. Pääbo himself has written an interesting autobiography, *Neanderthal Man*, in which he focuses more on his mother, the Estonian refugee and food chemist Karin Pääbo. When he was 13, she took him on a journey to Egypt, where he was seized by a deep fascination with mummies. That was how his career in research began.

He grew up in Bagarmossen, a suburb south of Stockholm, learned Russian at the Swedish Armed Forces' Language Institute (one of their most demanding courses) and went on to study Egyptology and Coptic at Uppsala University. A few years later, he abandoned Egyptology and switched to medical training. After four years of medicine, he began to conduct research in cell biology.

His day-to-day job involved studying a viral protein. In secret, he was also trying to isolate DNA from mummies thousands of years old. His boss knew nothing of this until Pääbo's first scientific article was well on the way to publication. The article appeared in an East German journal because the first mummies Pääbo had investigated came from a museum in East Berlin. It later transpired that what Pääbo had thought to be sensational DNA from a mummy was probably mostly the result of contamination by our contemporaries. But at least the idea was there. He had shown that DNA can survive in tissues several millennia old.

No one in the West took any notice of the East German publication. However, one year later Pääbo published another article, this time in the British journal *Nature*. This, by contrast, attracted great attention. One of the consequences was a letter from Allan Wilson of the University of Berkeley in California, one of the world's foremost specialists in evolution and DNA at that time. Wilson asked whether he could come and work in 'Professor' Pääbo's laboratory. Pääbo, who had just turned 30, was still working on his doctoral thesis, and he was far from having a lab to call his own. Instead, it was agreed that Pääbo would come and research at Wilson's laboratory in California, where Wilson and his colleagues had started to use DNA technology to unravel the early history of humankind.

By 1987 they were able to publish the first study using DNA technology to show that everyone living now ultimately originates from Africa. Our common foremother was a woman living in Africa some 200,000 years ago, who is usually known as 'mitochondrial Eve'. The name, of course, is a nod to the biblical creation myth. Mitochondria are discrete structures within cells that contain a small amount of DNA. Initially, all DNA analyses of the origins of humankind were based solely on mitochondrial DNA. This is because it is much easier to analyse than DNA from the cell nucleus (nuclear DNA), there being thousands of mitochondria in most cells, but only one nucleus. One limitation is that mitochondrial DNA can only be used to trace the origins of an individual in the maternal line, as we inherit our mitochondria solely from our mothers.

So by 1987, the technology available was already sophisticated enough to trace 'mitochondrial Eve', though the methods that were used appear unbelievably labour-intensive by today's standards. The young researchers from Wilson's laboratory visited maternity hospitals all over California to collect placentas from women from different parts of the world. Laboriously, they isolated the DNA from these placentas and calculated the results with their primitive computers.

Linda Vigilant, who is now married to Svante Pääbo, was one of Allan Wilson's postgraduate students. She conducted a follow-up study of mitochondrial Eve a few years later. The low-performance computers of the time were still so slow that it took a week to perform the calculations.

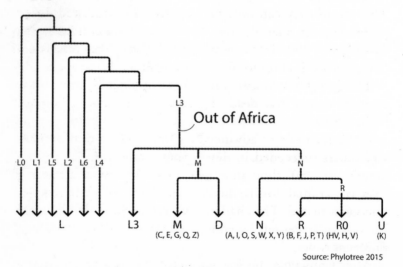

Source: Phylotree 2015

Figure 1 **Foremothers of Africa.** *'Eve', who lived in Africa around 200,000 years ago, became the foremother of everyone living today. About 60,000 years ago, one of the lines of her descendants left Africa and subsequently dispersed worldwide.*

But DNA technology had moved on considerably. A researcher in California had developed a technique for copying DNA, known as PCR (polymerase chain reaction), for which he was awarded the Nobel Prize a few years later. This method enabled researchers to work on single strands of hair, rather than having to use an entire placenta. Anthropologists from all over the world helped to supply strands of hair from a very wide range of populations. And the results of the first study were confirmed. About 200,000 years ago, there was a woman in Africa from whom everyone alive today is descended. She was thus our common foremother.

A few years later, in 1995, American researchers published their findings about mitochondrial Eve's male counterpart, known as 'Y-chromosome Adam'. Thanks to improved DNA technology and increasingly powerful computers, the Y chromosomes of men from all over the world could now be compared. Y chromosomes contain far more DNA than mitochondria, and they are passed on from father to son only. This means they can only be used to reconstruct the direct paternal line in a family tree. The results revealed that a man living about 200,000 years ago was the forefather of all men living today. Y-chromosome Adam lived in Africa too.

There was no longer any room for doubt. The multi-regional theory was dead. The origins of modern humans lie in Africa.

DNA technology advanced in leaps and bounds. Pääbo and others succeeded in developing methods for analysing samples several thousands of years old. By conducting tests on animal fossils, he learned how to keep samples uncontaminated. The challenge was to exclude any dust, old bacteria or traces left by people who had touched the fossils in recent times.

After California, he was offered a post as a professor at a zoological institution in Munich. There he impressed on his two doctoral students that they must avoid all forms of contamination, irradiate the whole laboratory with ultraviolet light every night, go straight to the special laboratory every morning without visiting any other laboratories with DNA samples, and take a whole range of other precautions.

In the summer of 1997 Pääbo published a DNA analysis of the world's most famous Neanderthal – the very skeleton that had been discovered in the Neander Valley in the mid-nineteenth century, giving Neanderthals their name. This time, the results were far more reliable than those obtained from the ancient mummy 12 years previously. The analysis, based on mitochondrial DNA, showed clearly that Neanderthals cannot be the forebears of modern Europeans. We are not their great-grandchildren; at least, we are not descended directly from them in the maternal line. It would be more accurate to compare our relationship with that of two groups of cousins whose common origins lie much further back in history.

This study had an enormous impact. Pääbo became a science superstar, particularly in Germany, where Neanderthals have had a unique status since they were first discovered in the nineteenth century.

***

Unfortunately, I have to admit I missed the news about this revolutionary study. I had started my job as *Dagens Nyheter*'s science editor just a few months previously and had not yet had time to acquaint myself with all the journals a science journalist needs to monitor. Nor was I yet on the right lists to receive advance warning of press conferences, and invitations to them.

But a few weeks later I met Pääbo at a seminar on the new gene technology, held in Oslo. We had dinner together, and my interest in this field of research was fired. Since then I have interviewed Pääbo many times, attended his lectures and reported in *Dagens Nyheter* on the publication of his studies in the world's leading scientific journals.

After the first mitochondrial analyses, he went on to analyse nuclear DNA, which, as noted above, is far trickier. However, a successful analysis helps provide a far fuller picture, as mitochondria contain no more than a few thousandths of a per cent of our total DNA. Moreover, they can only be passed on through the maternal line. The rest of our DNA, located in the cell nucleus, is inherited from both parents.

The first preliminary mapping of nuclear DNA published by Pääbo and his team in 2009 confirmed the picture that had emerged from the first analysis of mitochondrial DNA – that Neanderthals are *not* the forebears of modern humans. Rather, they should be seen as our genetic cousins.

But then came further input. More sophisticated analyses brought unexpected results. Pääbo himself was startled. In May 2010 he and his collaborators published an exhaustive analysis showing that Neanderthals and modern humans had in fact produced offspring. Features inherited from Neanderthals live on in us; the Neanderthals are not completely extinct.

Since then, the technology available for analysing prehistoric DNA has advanced even further, and it can now produce as high-definition an image as if researchers had examined you or me. The conclusions they have drawn can be summarised as follows.

It turns out that Neanderthals are our forebears after all, though they account for only about 2 per cent of our genetic make-up. When anatomically modern humans migrated from Africa to other parts of the world, they passed through the Middle East, including the region of today's Israel known as Galilee. The area was already inhabited by Neanderthals. They must have lived in parallel with anatomically modern humans in the same region for some time. People from the two groups had sex with each other on a number of occasions, producing offspring who were able to have healthy children themselves.

Modern-day people whose origins lie in Asia, Australia and the Americas have slightly more Neanderthal DNA than Europeans. Just over 2 per cent of their DNA is of Neanderthal origin, while an average European has just less than 2 per cent. This is probably the result of further interbreeding between Neanderthals and modern humans further east in Asia.

Moreover, the people who migrated eastwards into Asia, New Guinea and Australia appear to have interbred with another kind of archaic human called the Denisovans. Modern inhabitants of New Guinea have up to 6 per cent Denisovan DNA in their genetic make-up, and a smaller percentage can also be found in the Chinese population.

Characteristics inherited from Neanderthals can be detected even among Africans. This applies even to traditional groups such as the Yoruba of West Africa and the Mbuti people of the Democratic Republic of the Congo. However, the proportion is minute in such cases, and can be explained by the fact that some Europeans and Asians returned to Africa in the course of history.

There are commercial firms today selling DNA tests that, they claim, can reveal the percentage of Neanderthal or Denisovan DNA in an individual's genetic make-up. But Pääbo describes these tests as unreliable. The margin of error is so great that the result is next to meaningless. In retrospect, he regrets that his own team failed to apply for a patent on such tests; they would have been able to guarantee far higher standards.

Presumably it was an advantage for a young child living in the Middle East 54,000 years ago to have inherited certain features from Neanderthals. Such children may well have been healthier and more resistant to infection than others. Those of our forebears who migrated from Africa to the Middle East were part of a small group, possibly just a few dozen or a few hundred individuals. A few generations of breeding solely within the group had seriously depleted their immune defence. Inbreeding is bad for immune defence, so it is a good thing to have new, fresh blood from outside.

The American immunologist Peter Parham has identified a group of genes in our immune system that appear to have been inherited from Neanderthals. These genes must have helped the Troll Child and his peers to survive 54,000 years ago. Today, the same genes may be partly responsible for making the immune system too effective; as a result, it goes haywire in some cases, increasing the risk of autoimmune diseases such as multiple sclerosis and type 1 diabetes.

Researchers have identified two Neanderthal genes that affect the capacity to transform fat in the diet. In our time, bearing one of these two genes increases the risk of contracting type 2 diabetes. This disease is closely linked with being overweight, a major problem today. But things were different 54,000 years ago. For the first modern Europeans, it was an advantage for their bodies to be able to assimilate as much fat as possible; that reduced the risk of dying of starvation.

Svante Pääbo's research team has also identified a handful of other genes that seem to have been transferred from Neanderthals to modern human beings. They all affect a substance called keratin in our hair and skin. Both Asians and Europeans seem to have inherited variants of these keratin genes from the Neanderthals, but oddly enough, the sets of keratin genes concerned are different. It is not yet clear exactly how Neanderthal genes affect our hair and skin. If I were going to lay a bet, however, I would wager that straight hair is one of the characteristics inherited from the Neanderthals.

Pääbo himself prefers not to speculate about the appearance of Neanderthals' hair and skin. Today, other DNA researchers have developed genetic tests that provide clues about the colour of their skin, eyes and hair. A Spanish research team claims to have identified the hereditary disposition for red hair in one Neanderthal individual. Pääbo, however, thinks these tests are still far too unreliable for him to publish such information. During our interview I tried to press him on this point, arguing that the colour of Neanderthals' eyes, skin and hair would be sure to interest the general public. This kind of information would give a more vivid picture of what we experienced when we encountered Neanderthals.

Pääbo was not swayed by this argument. However, he did reveal something that has not yet been published: none of the Neanderthals he has analysed himself seem to have the genetic make-up associated with red hair. The Neanderthal individual that he has examined using the most high-resolution analysis, found in a Siberian cave, was probably dark-haired.

He was more comfortable talking about the approximately 87 gene variants that, though present in nearly all modern people, have not so far been found in any Neanderthals. The gene that has been researched in most detail is known as FOXP2.

There is a family living in Britain with several members of different generations who all have a serious language disability. All these individuals have a flaw in this particular gene, which clearly affects the ability to speak. This gene differs in just one small point between mice and chimpanzees. The difference between chimps and Neanderthals amounts to two points. In principle, modern humans and Neanderthals have identical FOXP2 genes. However, there is one tiny difference, discovered by Pääbo and his collaborators. It was difficult to find, being located well outside the gene itself, but it appears to have a significant impact on how the gene functions.

The Leipzig research group has developed special experimental mice bearing the human variant of the FOXP2 gene. Their squeaks are different from those of ordinary mice, and they have better memories. The difference lies in a particular kind of memory known to psychologists as 'procedural memory'. It involves the same mechanism we use when learning to ride a bike or dance; to begin with we have to think about each individual movement, but after a while we have internalised them. We have them at our fingertips – or rather in our cerebellum. We can make the movement automatically, without having to think about it. Learning to speak is no different.

It is reasonable to conclude that Neanderthal people could also talk to each other at some level – but not as we can.

***

Svante Pääbo tries to stick scrupulously to his scientific findings and to avoid speculation. However, there is a person who has not suffered from any such hang-ups when it comes to speculating about our encounters with Neanderthals – the American author Jean M. Auel. Her books, starting with *The Clan of the Cave Bear*, have sold in their millions. The first book in the series, which came out in 1980, describes how orphan Ayla is cared for by a different race of people. When Ayla grows up, she is repeatedly raped by the clan chief's son and gives birth to a son who is thus half Neanderthal, half modern human – this is known as a hybrid.

As regards hybridisation, you could say Auel was 30 years ahead of DNA research. When she wrote the book, there was no reliable scientific evidence of any such phenomenon. There were only individual skeletons, bones and teeth, regarded by some researchers as transitional forms.

Various other aspects of Auel's imaginative story are clearly mistaken. For example, she depicts the modern humans of our type as being blond and light-skinned, while the Neanderthals are described as being dark. But at the time the two met, things were probably the other way round. After all, anatomically modern humans had just left Africa, while the Neanderthals had developed in Europe and parts of Asia for several hundred millennia. Light skin improves people's chances of surviving in northerly latitudes.

Auel's descriptions of the Neanderthals' sign language and rigid social system are entertaining, and many of her inter-pretations are thought-provoking. She has clearly done her homework on archaeology, anthropology and botany. But we need to remember that her books are stories, not scientific works. Most of their content is pure fantasy.

One of the reasons why the books have sold in such enormous numbers, especially among teenagers, is the many sex scenes some of them contain. Although the descriptions are empathetic and very detailed, their tone is acceptably innocent. The name given to consensual sex is 'Pleasures'. But Auel's descriptions of sex acts between the heroine Ayla and the Neanderthal man are less pleasant. They involve a series of brutal rapes.

Over the last two years, when I told friends and acquaintances about the book I was writing, the question most frequently raised was what form sexual relations would actually have taken in practice. Most of those who allow themselves to speculate suggest the answer was rape, which tends to be my personal view. But such a scenario really provokes and upsets quite a few people. Some say there is absolutely no justification for speculating about issues that science can never settle conclusively. One of my closest friends accused me of taking a negative view of humankind. 'Maybe a Neanderthal boy and a modern girl fell in love with each other – that's just as feasible,' she commented.

Or, as a young female student said to her professor: 'If they decided to have children together, they must have known each other for quite some time.' Like so many other people, she based her reasoning on her own views on sex and morality.

The professor in question was Jean-Jacques Hublin, director of the department of human evolution at the Max Planck Institute for Evolutionary Anthropology in Leipzig. He is inclined to think that the encounters between Neanderthals and modern humans were over within a few minutes; his view is that they were acting in desperation.

Hublin's office is one floor down from Pääbo's, and it is also dominated by a Neanderthal. His specimen has a little more flesh on its bones; it is a white plaster bust displaying carefully reconstructed musculature. The bust was created back in the 1920s on the basis of the information then available, and in essentials it is still completely in line with the latest research.

Looking at the jutting, broad-nosed face impressed upon me even more just what a gulf there was between the Neanderthals and us. In my introductory story about the woman in Galilee who fell pregnant, I compared Neanderthals with trolls, the creatures that populate our old folk tales. This was not because I think it particularly likely that our troll myths originated in 40,000-year-old observations from real life. Rather, an encounter with a sizeable troll is something we can more or less imagine. The Neanderthals resembled us much more closely than today's chimpanzees do – but the difference was nonetheless considerable.

'I certainly don't think modern humans and Neanderthals got on with each other. I imagine they kept out of each other's way as much as possible,' says Hublin. His scientific discipline, palaeontology, involves attempting to reconstruct what happened thousands of years ago on the basis of fossil bone finds. Such research inevitably entails a degree of uncertainty. The world's foremost palaeontologists have somewhat different views on the differences between modern humans and Neanderthals and on the forms that encounters between the two groups took. Where opinions diverge, I have taken Hublin's line in this book. His position at the Max Planck Institute is unique. Working in close collaboration with DNA researchers, primatologists, archaeologists and anthropologists enables him to constantly update the information that can be gleaned from ancient bones.

The most ancient finds outside Africa of what are called 'anatomically modern humans' were discovered in Israel and are nearly 120,000 years old. But Hublin stresses that these archaic humans had not yet really developed into people like us. Presumably they died out without leaving any descendants. It was not until about 55,000 years later that a new wave of people migrated out of Africa (or possibly from the Arabian Peninsula). This time they were almost completely modern, with all that that implies in terms of skeletal shape and abilities. Israeli researchers have found a cranium from this group of modern humans in the Manot cave in western Galilee.

The first thing that happened when these modern humans left Africa was an encounter with Neanderthals. Neanderthal remains have been found a mere 40 kilometres (25 miles) east of the Manot cave, in the Amud cave in the hills above the Sea of Galilee.

Hublin stresses how thinly populated both Europe and Asia were at that time. The modern humans who had just arrived were not numerous, and the Neanderthals were under heavy pressure. The two groups can hardly have intermingled at watering holes; rather, they would have observed each other at a distance. They had sexual relations on a number of occasions. A small number of hybrid children were born and survived. A little later, the Neanderthals died out.

Hublin is convinced that they became extinct because we arrived on the scene. They died out first in the Middle East and later in the Caucasus, Siberia and Europe. We outcompeted the Neanderthals thanks to superior hunting methods and greater mobility – or we simply killed them off. Some researchers propose other possible explanations; for instance, we may have survived periods of cold and volcanic eruptions more successfully because we were better at making warm clothes out of pelts, using a needle and thread.

In Hublin's view, all these suggestions are excuses. The Neanderthals had survived ice ages and cold spells for hundreds of thousands of years. Sometimes their numbers fell drastically, but there was always a resurgence when the climate grew warmer – until we arrived. According to Hublin, the only reason for the alternative explanations is to avoid confronting us with the truth – that we actually wiped out an entire group of human beings. The Neanderthals had lived in Europe and Asia for several hundred thousand years. Then we came and took over. (There are at least two similar cases of eradication of other archaic humans when *Homo sapiens* later migrated further eastwards within Asia. Once we had arrived, both the Denisovans and the pygmy-like *Homo floresiensis* on the Indonesian island of Flores disappeared.)

While individual Neanderthals were physically stronger than us, we outdid them in other areas. Presumably we had superior powers of speech, as indicated by our slightly different FOXP2 genes. Language made it easier to maintain more extensive and cohesive networks. Finds of shells and rare stones show that some modern humans belonged to networks extending over 500 kilometres (310 miles), while Neanderthals exchanged artefacts over far shorter distances.

There are one or two finds suggesting that Neanderthals buried their dead. But chimpanzees, too, sometimes cover their dead with twigs and branches. The claim that Neanderthals laid flowers in graves, an assertion made on the basis of a find in Iraq, is hotly contested. In contrast, there is much incontrovertible evidence that modern humans went to considerable lengths to bury their dead and provide them with grave goods.

One other major and very clear difference is that modern humans played musical instruments and produced figurative art. There are a few simple patterns of lines that may have been incised by Neanderthals in Spain and by other, even earlier archaic humans on Java. But art representing animals, people and imaginary figures emerged only with modern humans.

The world's most ancient examples of musical instruments and figurative art have been found in Europe. The fact is that my forebears – in the direct maternal line – were among the anatomically modern, musical and artistic humans who first colonised Europe. This is revealed by DNA technology.

***

The year after my first meeting with Svante Pääbo, I went on a fact-finding trip to Iceland and visited deCODE Genetics in Reykjavik, where I interviewed another pioneer in DNA research, Kári Stefánsson. He was then in the process of setting up the genetic research company deCODE, based on two special sets of circumstances. Firstly, Iceland is an island where people have lived in relative isolation; secondly, many Icelanders have an interest in genealogy. Some of them can trace their ancestry as far back as the ninth century, when the island was first colonised.

At the time we met, Kári Stefánsson was a tall, fair-haired and strikingly elegant man in his fifties. Interviewing him was a curious experience, as other journalists have reported. It is his way to start by being singularly impertinent; then, if he thinks the journalist measures up, he changes tack, becoming friendly and opening up.

Fortunately I passed Kári Stefánsson's test. I got to spend a long time in his office, look at the fine maritime painting on the wall and listen to his plans. He gave a demonstration of his new software program, based on the exhaustive information that he had had a number of women copy from Iceland's church records. Tap a few keys, and you could see your own family tree unrolling centuries into the past, and study your relationship with other Icelandic families. That may sound trivial in the light of today's technology, but it was sensational in 1998.

DeCODE's researchers entered the information from these family trees in combination with DNA analyses from a large number of Icelanders. At the time, there was a heated debate on Iceland about the ethical and legal rules to be applied. One of the issues discussed was to what extent deCODE should have access to biological samples from the medical sphere. The debate calmed down eventually, and the company was granted extensive rights, but new legislation was also adopted that imposed certain limitations. Generally speaking, ethical regulations have become more stringent in most countries since the more Wild West-like conditions under which DNA research was conducted in the 1980s and 1990s.

Since the company's inception, Kári Stefánsson and deCODE's researchers have produced a great many scientific findings, particularly about certain gene variants that can increase or reduce the risk of particular diseases. In purely scientific terms, their work has been a real success story. They publish articles in major medical journals virtually every month. However, they have been less successful in business terms; though deCODE was supposed to be a commercial enterprise, it has never made a profit. In 2009 the firm went bankrupt and was taken over by interested parties from the US biotechnology sector.

Before that happened, deCODE began to sell genetic tests to private individuals on the internet, to make some money. I bought one of the tests, which was the most comprehensive that a private individual could purchase at that time. It covered a million 'items' in my DNA make-up and cost about SEK 15,000 (around £1,400). My reason for spending so much money was that I was doing research for my book *Vikten av gener* ('The Weight of Genes'), which came out in 2011.

I wanted to know how I would react to being told my risk of developing particular diseases.

In fact, all the medical information I received had very little impact on me. There was a slightly higher risk in some areas, and a slightly lower risk in others. For example, I have a slightly higher risk than average of developing certain types of cancer. One of them is skin cancer, which – being blonde and very fair-skinned – I could have worked out for myself.

I was also told I have the capacity to digest lactose, which means I can drink fresh milk – a common feature among northern Europeans, but one that is unusual in many parts of the world. That was hardly news to me; I had known for a long time that I can digest milk.

There was another piece of information that I found far more fascinating, something I could never have predicted. I discovered that I belong to haplogroup U5.

DeCODE's technicians had examined the DNA located in my mitochondria, the tiny structures within cells that you can only inherit from your mother, which are passed on virtually unchanged from mother to daughter over many generations. Sometimes mitochondrial DNA does undergo small changes – mutations – that make for slight variations between one individual and another. Such variations can be classified into genealogical trees. A single haplogroup corresponds to a particular branch of the tree that has a common origin. In other words, all the branches and sub-branches that share this origin have a common female ancestor. This enabled researchers back in the 1980s to trace the maternal lineage of the whole of humanity right back to our primeval foremother 'Eve', who lived in Africa some 200,000 years ago.

Now I learned that I was descended from one of 'Eve's daughters', who is known to researchers as U5. You might also say I am one of the daughters of 'Ursula', as group U5 is sometimes called. The name comes from a book published by the British geneticist Bryan Sykes in 2001 to popularise contemporary DNA research. His book was heavily criticised by other researchers even at the time of publication. There have been tremendous advances in research since then, so the book's contents are even less valid now than they were then. But since I like the name Ursula, I'll stick with it.

What the researchers were able to tell me when I got my results from deCODE was that the mutations typical of U5 were already present in Europe's early hunter population. That means that the woman we can call Ursula was an Ice Age hunter and belonged to the first people who colonised Europe.

I told two of my relatives with the same mitochondrial DNA about my results, and straight away their imaginations began to run riot. 'I've always liked to go rambling,' said one. 'You can tell we're descended from hunters,' said the other, who spends a good deal of free time hunting.

I explained that the fact that we belong to group U5 has very little − if any − impact on the traits we display. Mitochondria account for just a few thousandths of a per cent

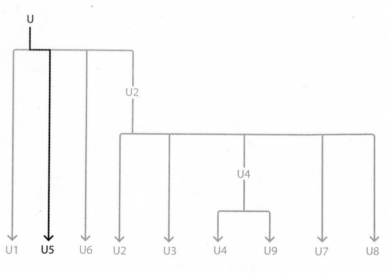

Source: van Oven 2009, Phylotree 2015

Figure 2 **U is spread by Ice Age hunters.** *One of the daughters of 'Eve' is known as U and probably lived in the Middle East. Her descendants spread throughout Ice Age Europe, and into Asia and North Africa.*

of our DNA, and the characteristics we inherit have been diluted many, many times since our foremother Ursula made her way into Europe.

It has more to do with feelings. We can think of our mother, grandmother, our grandmother's mother – and then imagine another thousand generations of daughters and mothers. In my case, that takes me back to the Ice Age people who played the world's oldest known flutes, who created the famous cave paintings, developed sewing needles and domesticated dogs.

I decided to travel and take a closer look at how my relatives lived.

CHAPTER THREE

The Flute Players

I arrived in the Swabian Jura on a sunny September day with the tree-clad slopes glowing in shades of yellow and red. My companion was a young archaeologist from the University of Tübingen.

At the entrance to the Hohle Fels cave we also met a local guide, Rainer Blumentritt, an elderly man who had discovered one of the local caves in his youth. Ever since that time, he has followed the work of the professional archaeologists at close quarters.

It was at just this time of year, in early autumn, that the first Europeans would have come here too. Here, in the limestone caves, they could seek shelter for the winter, and it was here that they could hunt reindeer when the animals had most meat on their bones and were migrating in great herds.

A few steps inside the mouth of the cave lies the chamber where these people lived. It looks like a cramped space, given that there were between 20 and 30 individuals in a group. They must have been crammed in together. But the young archaeologist accompanying me points out that it was probably warm and cosy as well. The fact that it was cramped made it easier to retain warmth during the cold winters of the Ice Age.

Archaeologists have dug out layer upon layer, excavating a total of several metres. They have found traces of Neanderthals right at the bottom, after which there are no human remains to be found for several centimetres. Above that are numerous traces of modern humans like you and me. The most ancient layer belongs to a culture that archaeologists know as the Aurignacian, identifiable by the way in which its people fashioned their stone tools.

According to the most recent efforts to date such remains, the people of the Aurignacian culture arrived in central Europe some 43,500 years ago. The first traces they left are

to be found in Austria, at an archaeological site called
Willendorf. The oldest finds in the Swabian Jura are nearly
as ancient.

Hohle Fels is not just a small cave where people once
lived. My companion has prepared me for viewing what he
calls 'the cathedral of the Ice Age'. And indeed, the inner
chamber does exceed all my expectations. The mountain
opens out into a vast chamber reminiscent of nothing so
much as the interior of a mediaeval church. Dim lamps glow
in niches along the walls. Though they are electric, I am sure
the light resembles that of the torches people used during the
Ice Age.

The young researcher from Tübingen and I walk around
in awe, admiring the great chamber. You can climb up onto
a shelf, as if in an amphitheatre. Then – quite unexpectedly –
we hear the limpid tones of a flute. The local guide has
switched on a recording relayed through loudspeakers. The
acoustics in the high-ceilinged chamber are extraordinary.
So this was what it sounded like when my forebears held their
ceremonies.

The flute we can hear in the recording is a reconstruction
fashioned out of mammoth ivory. You can still buy ivory from
extinct Siberian mammoths preserved in the permafrost – and
it's completely legal.

German archaeologists have found fragments of a total of
eight flutes in Hohle Fels and the nearby caves of
Geissenklösterle and Vogelherd. At a museum in nearby
Blaubeuren you can listen to the sound produced by reconstructions of these instruments. Four of the flutes are made
out of bones from birds' wings. They sound more or less
like someone blowing into a bottle – a very hollow sound.
Bones from swans' wings make higher-pitched sounds, while
those from vultures are larger, making for flutes with a
deeper pitch.

Making a flute out of the hollow bone of a bird may not
appear to be a very sophisticated process; all that is necessary
is to cut off the bone and drill holes at the appropriate
points. It is the four ivory flutes, which produce a sound

rivalling that of a silver flute in clarity and purity, that are truly mind-boggling. Carving a flute from a substance as hard as ivory called for great skill. You had to split a mammoth's tusk in half, scoop out the inside, and stick the two halves together again, forming a completely airtight join. The Aurignacians probably used resin as glue. Observing the positioning of the holes enables us to imagine what the music they produced would have sounded like.

Doubtless human beings have been singing and dancing since time immemorial, far longer than since the point when we first left the African continent. But the flutes from the Hohle Fels caves are the oldest musical instruments of which there is firm evidence. Researchers in Slovenia claim to have discovered an even more ancient bird bone with holes, which they believe to be a flute made by Neanderthals, but that find is surrounded by controversy. Nearly all other researchers are sceptical; they think the holes in the bone from Slovenia were produced by natural causes, perhaps by the bite of a hyena.

There is a consensus, however, that the Aurignacians in the Swabian Jura played the flute.

***

Not only did they have music in their lives, but they also created statuettes. Archaeologists working in the Swabian Jura have found around 50 figurines fashioned from ivory and stone.

The largest of these is the Lion Man from the Hohlenstein-Stadel cave, an ivory figure with a head like that of a lion but the body of a man. It was made nearly 40,000 years ago and is about 30 centimetres (12 inches) tall. A contemporary reconstruction shows that it must have taken about six weeks – for a skilled and experienced person working all day long in full daylight – to make.

In Hohle Fels, that great Ice Age cathedral, archaeologists have discovered a smaller lion man just 2.5 centimetres (one inch) tall that is more simply executed. Other finds from the same cave include a little ivory waterfowl, possibly a loon, and a particularly well-endowed female figure that is even older than the Lion Man and has been named 'the Venus of Hohle Fels'. She has enormous breasts, and her genitals are clearly marked by a deep cleft between her legs.

For a nomad constantly on the move between different dwelling places, it was practical for artistic artefacts to be small and easy to handle, and light enough to carry. We do not know exactly how they were used, but it requires no great leap of the imagination to see the connections between these art objects and shamanistic beliefs studied by ethnologists over the last 150 years among Swedish Sami, Siberian nomads, Native Americans and the South African Bushmen (also known as the San people).

The cave lion was the largest and most dangerous predator that posed a threat to our forebears in the Swabian Jura. They certainly both feared it and admired its strength and speed. There are many descriptions of how shamans take on the shape of an animal in their ceremonies, often using masks to do so. It is not much of a stretch of the imagination to see the cave lion as part of this repertoire. Waterfowl, too, often play a special role in a shamanistic world view. This is because of their ability to make their way freely between the three levels of the universe; they can fly and thus reach the heavens, they can walk the earth like humans, and they can dive into the depths to reach the underworld.

The large-breasted female figures appear in both Europe and Siberia throughout the nearly 30,000 years of the Ice Age. They may have had a place in female rites around fertility and giving birth. That, at any rate, is the view of Jill Cook, a curator of Stone Age art at London's British Museum.

Nicholas Conard, a professor of archaeology at the University of Tübingen who is in overall charge of the excavations at Hohle Fels, takes a somewhat different view. He prefers to extend the significance of these female figures to encompass fecundity in general, including the whole of nature and the wild animals that were people's quarry. He has suggested that Angela Merkel, the Christian Democrat German chancellor, should wear a replica of the Venus of

Hohle Fels around her neck. She has demurred so far; perhaps the figurine's clearly defined genitalia are a little too provocative for the circles in which she moves.

Some of the art objects from the Swabian Jura are more ancient than any other figurative art anywhere in the world. There are, of course, non-figurative motifs that are tens of thousands of years older than the objects found in Swabia, such as the zigzag patterns on stones from Blombos in South Africa, which have been dated as at least 75,000 years old. But the figures from the caves in southern Germany show that over 40,000 years ago people in Europe were able to depict both creatures that actually exist in real life, such as waterfowl, and imaginary beings, such as the lion men. They must have had mental powers just like ours today.

Conard has advanced the rather provocative proposition that this figurative art first emerged when humans reached the region that is now Germany. He hypothesises that new challenges, such as a cold climate and rivalry with groups already living there, triggered abilities that had hitherto lain dormant. He concedes that people were discovering figurative art in other places too, but argues that those in Swabia were the first to do so.

Conard is aware that his hypothesis is a provocative one. He says he is quite ready to abandon it as soon as archaeologists in other countries find indisputable proof of figurative art or musical instruments that predate the Swabian finds. But the reason why the oldest figurative art and the oldest musical instruments have been found in Germany may be that archaeologists have spent more time searching there than in most other places on earth. That was my own view when I met Conard in Tübingen in the autumn of 2013.

And one year after my visit, strong indications emerged that he is mistaken. An international team of researchers published new information dating cave paintings from the Indonesian island of Sulawesi. The most ancient paintings, it transpired, are some 40,000 years old – about as early as the very oldest paintings in European caves, and nearly as old as the oldest statuettes from Swabia.

When I asked Conard to comment on the new dates, he conceded that people on the other side of the earth were able to create art just as early as those dwelling in the caves of Swabia. However, he claims that the art found in Indonesia and that of Swabia may have developed in parallel, each quite independently of the other. He is now quite alone among the world's leading experts in human evolution in holding this view. Most believe that art and music were already part of our baggage when we left Africa some 55,000 years ago. Culture has followed us on our wanderings ever since, both eastwards and westwards.

I am nonetheless convinced that Conard has an important point in viewing art and music as playing a decisive role in our early ancestors' ability to stay together and survive. But I believe that was the case in Africa, Europe and elsewhere. Creativity and artistic ability have been so crucial to our survival that these abilities are written into our genes – despite the dark downside.

**

Self-expression through art, music and storytelling is one of the main forces that drive human beings. That was so during the Ice Age, and it remains so today. Yet some of us pay a high price for the continued existence of these features in the population.

Talents such as being able to paint, play music and tell stories are in part hereditary. Families in which such gifts are common also tend to be affected by psychotic conditions such as schizophrenia and bipolar disorder. Both these conditions affect about 1 per cent of the population worldwide, in all the countries that have been studied. They can be very disabling. Relatives of people affected by schizophrenia are often successful as artists and musicians. Unfortunately, that does not apply to the individuals suffering from the condition, as it limits their capabilities to such an extent. People with bipolar disorder more often succeed in creative occupations themselves, and the same is true of close family members.

One of the first people to observe this correlation was an Icelandic researcher called Jon Love Karlsson who published a groundbreaking study back in 1970. His methods leave something to be desired by contemporary scientific standards. He based his research on Iceland's well-documented genealogical information, which – as mentioned in the previous chapter – is probably more comprehensive than anywhere else in the world. This information was compared with the records of patients at Reykjavik's mental hospital and the Icelandic version of *Who's Who*.

Karlsson believed that the risk of schizophrenia was governed by just two genes. Today's DNA research has shown that it actually depends on hundreds of genes, in conjunction with unknown environmental factors. But the link between creativity and psychotic disorders has been confirmed, and that applies to the most recent, largest and best conducted studies.

Many of the gene variants involved emerge as entirely new mutations when a new life comes into being. This is cutting-edge knowledge that has revolutionised research into schizophrenia and a number of other psychotic disorders. The same calculations of new mutations also play a decisive role when DNA is used in genealogy and research into the history of humankind. If you know how many mutations accompany each new baby, you can work out the speed of human evolution. You can set the stopwatch – and time the rate of evolution.

The first major breakthrough in this area again came from Iceland. Thanks to the new technology for DNA analysis, a team of Icelandic researchers were able in 2012 to compare parents and children in 78 families. Their comparisons enabled them to confirm a theory that the children of fathers who were older at the time of conception have a higher risk of schizophrenia. The reason for this is that sperm from older fathers contain more mutations. An equally groundbreaking aspect was that the Icelanders were able to quantify the number of new mutations each baby is born with. They average about 30 on the mother's side and about 30 on the father's side if he is in his thirties. If he is in his sixties, the number of mutations on his side rises to about 60.

Svante Pääbo's team of researchers in Leipzig has attempted to calculate the rate of mutation using a completely different method. They have compared DNA from ancient skeletons with DNA from people alive today. In the past, it was difficult to reconcile the findings from Iceland with estimates of how DNA changes in the course of evolution. The different methods resulted in totally different figures. You could say that the researchers' watches seemed to be ticking at different speeds. But in the autumn of 2014, Pääbo's team published analyses of the DNA of a 45,000-year-old man from a place called Ust'-Ishim in western Siberia. He was the oldest modern human ever subjected to DNA analysis and therefore provided the best benchmark for calibrating the DNA watch.

A degree of uncertainty remains, partly because we do not know exactly how old our forefathers and foremothers were when they had their children. But now the results of the two methods are far better matched. We can assume that each new individual born inherits about 30 new mutations from each parent. If that individual is unlucky, these new mutations can end up in unfavourable locations, where they may be partly responsible for disorders such as schizophrenia and manic depression.

Completely new mutations thus account for part of the risk of contracting one of these psychotic disorders. But there are also many gene variants that are clearly congenital, inherited from parents, maternal and paternal grandparents, and previous generations and lineages. Such gene variants have attached themselves to the human race, even though psychotic disorders bring such huge disadvantages with them. They reduce the chances of survival, and schizophrenia, at any rate, reduces the likelihood of having children of one's own.

The only possible explanation for the continued existence of this inherited trait is that it also confers some advantages. For those who have mental illness in their family, it may be a comfort to reflect that a predisposition to psychosis is a double-edged thing. Such disorders are undoubtedly disabling; they often place a heavy burden on both the sufferer and on his or her loved ones. Yet they also represent a gift – in the form of creativity and extra energy. These characteristics have been tremendous assets in the course of human history. Indeed, they have actually proven to be indispensable to our development.

That's how I tend to look at it.

CHAPTER FOUR

First on the Scene in Europe

Over 42,000 years ago, the Aurignacians were already migrating across large areas of Europe. Some of the oldest and best finds come from an archaeological site in Russia, Kostenki, which lies 400 kilometres (250 miles) south of Moscow, on the River Don. Researchers have analysed DNA from the skeleton of a young man found here. Known as Kostenki-14, he lived about 38,000 years ago. He died in his twenties and was buried in a foetal position, with generous amounts of red ochre scattered over his body.

His mitochondrial DNA belongs to group U2, which is closely related to my own haplogroup, U5. Although U2 is very unusual today, it still crops up throughout Europe, as well as in central Asia and India. More exhaustive analysis of his mitochondrial DNA shows he was related to contemporary Europeans. He confirms the view that a group of people from Africa arrived in the Middle East about 55,000 years ago. They intermingled with a small number of Neanderthals, and shortly after this interbreeding the group split up. Some of them went further east, becoming South-east Asians and Australians in the fullness of time. Some remained in the Middle East and the Caucasus, while others began to migrate towards Europe. The man from Kostenki clearly belonged to the European group.

His grave, containing a well-preserved skeleton, was discovered back in the 1950s. Russian researchers tried to reconstruct his appearance on the basis of his bones. There are several copies of the reconstruction at museums in Moscow, Saint Petersburg and the rural locality of Kostenki.

These models show a slender young man with a broad but straight nose, unusually pronounced eyebrow arches and full lips. He was just 1.6 metres (5¼ feet) in height, with a strikingly small cranium. His teeth are broad and somewhat worn, though otherwise healthy and in good condition, with a narrow gap between the front teeth in his upper jaw. The models depict him as dark-skinned, with dark, curly hair. Although that is certainly a reasonable assumption, researchers have not yet published any such details in the DNA analyses that are publicly available.

Nearby finds from the same period show that wild horses were the most common prey of the Kostenki man and his contemporaries. Archaeologists also found some hare and mammoth bones in his grave. At that time, clumps of trees, mainly willows, were beginning to grow in the steppes again, as the climate had recently become warmer after one of the Ice Age's colder periods. Life had become easier again after the worst catastrophe that Europe had experienced in 100,000 years.

The waxing and waning of ice ages is mainly the result of cyclical changes in the earth's orbit around the sun. The earth's axis can be tilted to a greater or lesser degree, it can be tilted in different directions, and the elliptical shape of the earth's orbit can vary. In certain positions, when the northern latitudes receive particularly little sunlight, a new global ice age can be triggered. These shifts are known as Milankovitch cycles, after the Serbian physicist who first described the pattern.

But there are other factors that can affect the climate at both global and regional level. Volcanoes, for instance, can discharge ash that filters out the sun's rays, thereby causing the earth to cool down.

Some 39,000 years ago there was a really 'Big Bang' – a volcanic eruption in the area around Naples, Italy. Magma poured out in quantities about 80 times greater than at the better-known eruption that buried the Roman town of Pompeii in AD 79. The plume from the magma rose 40 kilometres (25 miles) into the atmosphere. The ash spread mainly eastwards, over Greece and Bulgaria, the Black Sea and Russia. In Kostenki, archaeologists can clearly distinguish a thick layer composed mainly of dust, which helps them to date their finds. It seems likely that the sky was darkened for a number of years, that the climate grew considerably colder, and that the ground was covered in decimetre-thick layers of ash – so thick in many places that it was impossible for animals to graze.

Some researchers believe that this volcanic eruption sounded the death knell for the Neanderthals. The Neanderthals left many traces that go back beyond 39,000 years, but there are scarcely any more recent ones that can definitely be attributed to them. The hypothesis is that at least some of us modern humans, unlike the Neanderthals, managed to adapt to the newer, harsher conditions after the 'Big Bang'. Some researchers believe, for instance, that we began to use bone needles in this very period. This revolutionary new technology, the claim goes, enabled us to make warm, weatherproof clothes out of animal hides, and thereby to survive the most bitterly cold years.

The oldest known finds of eyed needles come from Kostenki and two other archaeological sites in Russia: Mezmaiskaya, in the northern Caucasus, and a location in the Altai Mountains of southern Siberia. They are estimated to be between 35,000 and 40,000 years old.

Clearly, needles and warm, well-made clothes may have helped us to survive. But they are probably not the whole explanation for why we survived while the Neanderthals died out. We need to remember that the Neanderthals had lived through many harsh periods of cold in the course of several ice ages. They recovered each time. It was not until we modern humans came on the scene that they were doomed to perish. The pattern is recognisable in many places in the Middle East, the Caucasus, Siberia and Europe. The Neanderthals lived there for thousands of years, then we arrived and the Neanderthals disappeared. In some places they had already vanished before our arrival. That can be seen, for instance, in the layers excavated at Hohle Fels. Excavations at other sites bear witness to an overlap of several thousand years, during which both groups may have coincided – albeit at a decent distance from each other. There are also sites where all traces of Neanderthals stop abruptly, followed immediately by remains left by modern humans.

***

Very close to the grave of the Kostenki-14 man, in layers dating back to the same period, archaeologists have found needles along with a number of objects typical of the Aurignacian culture. There are tools made of bone and antler, and stones taken from rocky outcrops 1,500 kilometres (930 miles) away.

In addition, the Aurignacians of Kostenki made jewellery using the canine teeth of Arctic foxes and shells all the way from the Black Sea, 500 kilometres (310 miles) to the south. They also fashioned tubular beads with grooved spiral patterns out of the bones of foxes and birds.

Similar beads have been found throughout the region that has yielded Aurignacian finds. The site excavated at Abri Castanet in the French Dordogne was nothing less than a factory, where people mass-produced beads from mammoth tusks, reindeer antlers – and soapstone. Since that type of stone did not occur locally, they must have brought it from the Pyrenees, several dozen kilometres further south. The people in the hills of the Dordogne also adorned themselves with shells from both the Mediterranean and the Atlantic seaboard. Either they covered the whole distance on foot – up to 200 kilometres (125 miles) – or else they had extensive networks enabling them to barter goods with other groups.

The Aurignacians probably travelled from the Middle East through the region that is now Turkey. In any event, they travelled westward along the Danube just over 43,000 years ago. We can set aside the question of whether they had flutes and art objects made of ivory among their scant possessions, though I believe that to be the case. It is clear that their garments were embellished with various adornments.

Yet there were also anatomically modern humans – people like us – in Europe at an even earlier period. The very oldest remains at Kostenki are assumed to be of Neanderthal origin. But the objects dating from 45,000 years ago – and earlier – appear to come from modern humans. In other words, those layers are significantly older than both the Kostenki-14 man and all the finds ascribed to the Aurignacian culture.

Stone tools of at least equal antiquity – apparently produced by modern humans – have been discovered at a number of sites in Hungary and the Czech Republic. It would seem that small groups of our type of humans made incursions into Europe at a very early stage, perhaps starting even earlier than 50,000 years ago. But these early pioneers did not survive. It was not until the Aurignacians that Europe acquired a viable population of modern humans.

There are also a number of finds from Italy and Greece that are now generally attributed to modern humans. The culture to which they belong, known as the Uluzzian, was discovered in the 1960s. For many years the view prevailed that these tools and adornments belonged to a group of unusually sophisticated Neanderthals. The stone tools appear to show a curious blend of the Neanderthals' production methods and those of modern humans. These finds also include shells and teeth perforated for use as pendants, remnants of red ochre, and tools made of bone. Just a few years ago, Italian researchers analysed two milk teeth from Italy's Grotta del Cavallo. The shape of these teeth has now convinced many – though not all – experts that they in fact belonged to an anatomically modern human. The debate continues; there are no DNA analyses.

All traces of the Uluzzians came to an end around 39,300 years ago. There is most probably a connection with the major volcanic eruption that took place at that time, very close by. But before the Uluzzians disappeared, they managed to have a decisive impact on their surroundings. They – or other pioneering groups of modern humans – introduced new customs into Europe.

The Neanderthals were not slow to copy them. There was another noteworthy culture in western Europe that appears to be a hybrid between the respective cultures of the Neanderthals and modern humans. Known as the Châtelperronian, it has been identified in northern Spain and south-west France. The people belonging to this culture seem to have buried their dead on occasion, if only in a simple way, and they seem to have used personal adornments, arrows and, to some extent, coloured pigments.

Researchers have argued a great deal about who lay behind the Châtelperronian culture, but now, thanks to new and more sophisticated methods of radiocarbon dating, a picture is finally beginning to emerge. Everything suggests that the Châtelperronians were Neanderthals who imitated modern humans. Inspired by new arrivals in the region, they began to use personal adornments, make-up and spears.

The new and more precise radiocarbon dating was carried out at Oxford University under the leadership of Tom Higham. It suggests that all the Neanderthals of Europe disappeared at least 39,000 years ago; at any rate, there is no definite evidence of Neanderthals that is any more recent than that. But the new results of dating also show that Neanderthals and modern humans must have coexisted in Europe for thousands of years. That means there was ample time for the Neanderthals to acquire innovations from modern humans.

As we have seen, Jean-Jacques Hublin, the palaeontologist from Leipzig, is convinced that the two groups were suspicious of one another and that they kept their distance as much as possible. But he also suggests that they would sometimes have observed each other at arm's length. This would have enabled the Neanderthals to see that the modern humans used spears, which they threw at their quarry – a brilliant invention that made hunting both safer and more effective. The older method was to run after your prey and spear it directly, a technique that Neanderthals had used for several hundreds of thousands of years. It meant risking your life, of course, but they had no better method.

The Châtelperronian finds show that they suddenly started to produce spears at the very time when modern human beings first arrived in Europe. Neanderthals' spears were very similar to those of modern humans and could be used in exactly the same way, but the two groups made their stone tools in slightly different ways. This fact strengthens Hublin in his conviction that the Neanderthals copied modern humans at a distance. In his view, there was no question of the two groups fraternising with each other. He may, however, be able to concede that they bartered with each other on rare occasions. That would explain why the layers with Neanderthal artefacts contain beads similar to those of modern humans.

***

The new ways adopted by the Neanderthals may have enabled them to survive for slightly longer; nonetheless, their downfall was a foregone conclusion once humans of our type began to take over Europe.

One of many attempts to account for the fact that we survived and they died out is based on the claim that we modern humans were supposedly less finicky about food. It was claimed that we were better at eating vegetables, such as roots rich in starch. But new research, including that of Amanda Henry at the Max Planck Institute in Leipzig, puts paid to that explanation. Having studied microscopic deposits on fossil teeth, she can bear witness to the fact that the Neanderthals also consumed a great deal of plant starch – so it is false to claim that they were very one-sided carnivores and died out for that reason.

On the other hand, we may have been better at catching fish and small, fast-moving animals such as hares and birds. It is conceivable that we were more skilled at making fishing nets out of plant fibres. Being able to fish and hunt using such nets was a major advantage. Firstly, it meant there was a more varied basis for people's diet, which was thus more reliable. If there were no big animals to hunt, they could always go down to the river and catch a few fish. Secondly, it meant more members of the group could contribute their labour. Hunting large mammals could often be dangerous and physically demanding; only strong, healthy people could manage it. But laying and emptying fishing nets and traps was feasible even for weaker, older and disabled people. The art of fishing and hunting hares and birds may well have played a decisive role in securing our survival in Europe.

Maybe we outcompeted the Neanderthals by having more effective hunting methods. Maybe we simply killed them whenever the opportunity presented itself. I think Hublin is right in ascribing limited significance to climate change and sewing needles, but all the more to language, art, music and more extensive social networks.

However, the fact that the Neanderthals were already subject to severe pressures in the late Ice Age, before we arrived, must have simplified matters. A number of DNA studies show that some of them were very inbred towards the end. A Neanderthal boy in the Denisova cave in southern Siberia had so little genetic variation that his parents must have been half-siblings or the equivalent, while their parents were themselves the result of many generations of inbreeding within a small, restricted group. A Swedish study shows that towards the end there can only have been a few thousand Neanderthals in the whole of Europe. Their population dwindled rapidly about 50,000 years ago, as can be seen from the fact that their DNA became less and less varied.

The arrival of modern humans, the period of cold and the volcanic eruption in Italy would thus have sounded the death knell for a group that was already extremely vulnerable.

In the past, some researchers insisted that the Neanderthals had ample intellectual capacity to develop their new culture in a wholly independent way. But such arguments are heard less and less these days. It is abundantly clear that the Neanderthals lived in Europe in very much the same way for several hundred millennia, but then suddenly changed their way of life as soon as humans like us turned up.

In her bestselling novels, Jean M. Auel describes how Ayla, a Stone Age girl belonging to a group of modern humans, grows up among Neanderthals. We cannot rule out the possibility that something like that may have happened in real life. There may have been individuals who moved between the two groups, bringing knowledge and traditions with them. After all, the genetic evidence shows that Neanderthals and modern humans interbred in the Middle East.

They also had hybrid offspring in Europe. For a long time, a number of anthropologists have claimed that certain excavated skeletons show clear anatomical features from both Neanderthals and modern humans. For instance, there are the two finds from the Peştera cu Oase cave in Romania: the cranium of a 15-year-old and the lower jaw of an adult. The latter specimen has been radiocarbon-dated, and its owner is estimated to have lived about 40,000 years ago.

In the spring of 2015, Svante Pääbo and his colleagues managed to prove that the lower jaw really does have a significant proportion of Neanderthal DNA – between 5 and 11 per cent of the genetic material. And the Neanderthal input had clearly come a mere four or five generations ago, as the pieces of DNA are in such long, unmixed sequences. That implies that the great-great-grandfather of the individual from Peştera cu Oase – or a relative at a similar distance in the ascending line – was a Neanderthal.

But people living in present-day Europe have no trace of this in our genetic material. In other words, when modern humans in Europe interbred with Neanderthals, their descendants must have died out. Researchers can only see definite traces of interbreeding somewhere in the Middle East about 54,000 years ago, and further east in Asia.

Neanderthals had bigger brains than ours, and they were by no means stupid. They were skilled hunters, and as regards dexterity they could give us a run for our money in many respects. Their stone tools were symmetrical and functional, and I have heard many archaeologists say how very difficult it is to learn how to make such tools. Clearly the Neanderthals also had the capacity to develop their technology, even if they did so by copying us.

But there is no credible evidence today to suggest that they made any use of art or musical instruments. They were probably unable to think in symbols to the same extent as we can. It is very clear that they had a feeling for symmetry, but not for what we consider to be aesthetics and art.

I had a eureka moment when I heard Jean-Jacques Hublin mention the difference between symmetry and aesthetics. I thought of the time when I used to work in a cake shop as a teenager and learned how to decorate cakes and gateaux. Nearly all beginners make the same mistake: they try to make the patterns completely symmetrical. Once I had plucked up the courage to get away from symmetry, the gateaux I decorated looked a lot more attractive. Defying symmetry − just suppose that is one of the keys to what is uniquely human?

Like Hublin, I believe we modern humans bore some of the responsibility for the demise of the Neanderthals. We can reflect on the moral guilt that that might imply, even if it has − hopefully − passed the statute of limitations by now. After all, the Neanderthals were a kind of human being, even if they were not quite like us. Is their disappearance comparable with driving a species of animal to extinction? Or would it be more accurate to regard it as genocide?

Whatever the conclusion, we should be very careful not to look down on the Neanderthals. They managed to inhabit Europe for much longer than we have done so far. They were here for at least a few hundred millennia; indeed, if you also count their forerunners − sometimes called *Homo heidelbergensis* − for over 400,000 years.

The Aurignacian culture dominated Europe for about 10,000 years. It, too, lasted for much longer than any realm we know of from historical time. But then a new wave of migrants arrived on the scene.

CHAPTER FIVE

Mammoths in Brno

The Czech city of Brno is classic ground for anyone with an interest in genetics. I have been here once before to write about the monk Gregor Mendel. It was he who showed in the 1860s that hereditary traits are passed on in discrete units, which we today know as genes. The monastery where Mendel worked still exists, having been refurbished after years of decline during the communist era. Monasteries were less than popular at that time. Biology, too, was regarded with great suspicion – particularly when Joseph Stalin held sway in the Soviet Union. Genetics was a particular taboo, held to be bourgeois and counter-revolutionary.

Now I am in Brno for the second time, I take the opportunity to have another look at the Augustinian monastery and museum that present Mendel's life and work. The curly tendrils of a few pea plants in bloom are to be seen outside the entrance. Peas were the plants Mendel focused on, being easy and practical to work with. Cultivation was followed by calculation, time and again: yellow peas, green peas, red flowers, white flowers, tall plants, short plants, and so on. On the basis of seven characteristics he identified in peas, Mendel formulated the laws of heredity and described dominant and recessive traits.

It took over 40 years for Mendel's findings to reach the world outside Brno. But once they started to be applied, they revolutionised the breeding of plants and domestic animals, and virtually the whole of biology. Unfortunately, his findings have also been misinterpreted. Today we know that heredity rarely works as simply as with Mendel's green and yellow peas. Most characteristics are considerably more complex; they are affected by many different genes, as well as by environmental factors.

After visiting the Mendel Museum, I take the tram a few stops further on. This time, I've come to Brno to learn more about the great European Ice Age culture that followed the Aurignacian culture, known as the Gravettian.

Outside the city are a number of Europe's most important Ice Age sites, the most famous of which is Dolní Věstonice. Many of the finds from these sites are on display at the Anthropos Pavilion on the outskirts of Brno. On the tram, I attempt to communicate with a group of Czech ladies in late middle age, to find out where I need to get off. English doesn't work, so I resort to my extremely basic Russian. When they finally grasp what it is I want to know, they exclaim, 'Aha! *Mamut!*' One of them gets off at the same stop, not because she herself has any business there, but just to point me in the right direction. On arrival, I see what they were referring to. A gigantic woolly mammoth dominates the museum, two storeys high. The calf alongside the full-grown mammoth is nearly the size of a cow.

In the upper level I view the famous triple grave of three young people, all of whom died in their teens or twenties. In the middle lies a crippled woman whose skeleton displays apparently congenital deformities. She is on her back, with a man on either side of her. One of them lies face down next to the woman, his arm entwined with hers. The other is a little further away, but his hand is outstretched over the woman's pubis. Their heads are sprinkled with ochre, as is the woman's crotch and the man's hand.

Two of these individuals can be described as my relatives. Their mitochondrial DNA belongs to group U5, just like mine. But their U5 is a very early variant that does not correspond to any that exists today. The third individual – the one whose hand rests on the woman's pubic area – belongs to group U8.

One interpretation is that the man and the woman belonging to group U5 are siblings, while the man from group U8 is the woman's partner. Brother and sister lie alongside one another, their arms entwined. The woman's partner lies a little further away, but his hand rests on her genital area. Their position in the grave reflects their relationship in life.

The young people in the triple grave lived approximately 31,000 years ago. But the oldest finds at Dolní Věstonice go back all of 34,000 years, according to the most recent datings. All of them have been assigned to the Gravettian culture. The tools and art objects of the Aurignacian and Gravettian cultures differ quite markedly.

At one of the city's universities I meet Jiří Svoboda, the Brno archaeologist in charge of the excavations at Dolní Věstonice. A quiet man in his sixties, he is among Europe's most respected archaeologists. Svoboda is convinced that the Aurignacian and Gravettian cultures represent two distinct waves of migration into Europe. The appearance of the artefacts indicates that the people of the Gravettian came from the south, from the Middle East and Mediterranean coastal areas.

The only individual belonging to the Aurignacian culture whose DNA has been analysed is the Kostenki-14 man from Russia, who belonged to the U2 haplogroup. The three Gravettian individuals from Dolní Věstonice belong to the U5 and U8 groups. While that circumstance provides rather little DNA evidence for Svoboda's theory of two separate migratory waves, there is no counter-evidence. He may well be right, and analyses of fossil DNA may provide more evidence in his favour in the long run.

If the theory is correct, the Aurignacian and Gravettian cultures both originated in the Middle East some 50,000 years ago and arrived in Europe in two separate waves. The group from which both originated included 'Ursula', a woman from haplogroup U who became the foremother of all the individuals falling into a total of nine subgroups: U2, U4, U5, U8, and so on. People belonging to the original group in the Middle East made repeated attempts to migrate into Europe. However, just two waves of migration during the Ice Age had a permanent impact: the Aurignacian culture, which arrived here over 43,000 years ago, and the Gravettian culture, which appeared some 34,000 years ago.

The rich finds from Dolní Věstonice provide a huge amount of information about life in central Europe between 34,000 and 20,000 years ago. The evidence shows how the people regularly returned to their settlements. Some of them may have lived permanently at Dolní Věstonice. They were highly specialised in hunting mammoths, but their diet also included hares and birds.

There are a number of finds of personal ornaments, such as perforated foxes' teeth and beads, but fewer than is usual in remains from the Aurignacian culture. The Gravettians seem to have focused on decorating their headgear.

A number of art objects have been preserved. The most famous of these is the 'Venus of Dolní Věstonice', a large-breasted female figure made of fired clay. She is several thousand years older than the oldest ceramic vessels we know of, which were found in Japan and China.

The people of Dolní Věstonice were nomads who had no desire to lug heavy ceramic vessels around, Svoboda explains. They cooked their food in containers made of hide, heating the water with hot stones taken from the fire. That was the Stone Age version of our kettles – and it was a surprisingly effective method.

Instead, they used clay to make miniature figures of both animals and people, which archaeologists have discovered in large numbers. Many of them are broken, and the vestiges have been found directly next to hearths. It looks as if the inhabitants of Dolní Věstonice moulded objects out of clay, then put them on the fire before they had had time to dry out completely. That made the moisture in the clay expand, and the object explode like popcorn. We can only speculate about whether exploding clay figures were a party piece just for entertainment, or whether they were part of some rite.

Another notable occupation in Dolní Věstonice seems to have been bashing each other over the head. It looks as if clubs or other hard objects were involved. Many skulls show signs of serious injuries, though they healed before the individuals concerned died of some other cause.

One of the questions hotly disputed by researchers is whether the people of Dolní Věstonice kept domesticated dogs. The excavations have revealed many wolf bones, some of them from unusually small wolves. Many archaeologists interpret these small wolf skeletons as a sign that they were more probably tame dogs.

Svoboda is diplomatic when discussing this subject. He works with a number of other researchers whose views on the matter vary considerably. However, he draws attention to a factor that may be decisive: though there are large numbers of bones from animal prey in the settlements, none of them appear to have been gnawed by dogs. What that suggests is that wolves approached human settlements because they could readily gain access to meat there. But they remained on the periphery and did not become our tame companions until later. I shall return later to the heated debate about the first dogs.

The Venus of Dolní Věstonice is quite corpulent, as are many other female figures of which archaeologists have found fragments. They have so much subcutaneous fat that the skin on their backs falls into folds. This is something of a paradox, as bones and teeth reveal that the people of Dolní Věstonice suffered periodically from famine. Life could be hard on the plains of central Europe, with huge temperature swings. Finally, about 20,000 years ago, it grew so cold that people could no longer remain there. This was the beginning of the very coldest period of the Ice Age.

Central Europe was no longer habitable. The mammoth hunters of Dolní Věstonice moved to regions where the climate was milder and life more bearable. As for me, I now travel down to south-west France and Spain. I have reason to believe that some of my forebears made their way to that region.

CHAPTER SIX

Cro-Magnon

I travel to Les Eyzies-de-Tayac, and find lodgings at Cro-Magnon itself. Les Eyzies lies in the hills of the Dordogne, in south-west France, a few hours' journey east of Bordeaux by local train. For anyone interested in studying Europe's Ice Age, this is the centre of the world. Tourists come in their hundreds of thousands to view archaeological sites, cave paintings and museums. The little village is well set up to accommodate all these Ice Age tourists.

Everything began with Cro-Magnon. A well-to-do local farmer, Magnon by name, planned to build a road across his land to the new railway station. His labourers gathered stones for the road from a rock shelter (*abri* in standard French, *cro* in the regional dialect). Such shelters are a common geological phenomenon in these hilly limestone regions. The hillside is hollowed out by groundwater percolating through and by frost erosion, leaving a sheltered space with a natural roof.

Several human skeletons of ancient appearance were discovered under the Cro-Magnon rock shelter. This happened in 1868, just a few years after the discovery of the Neanderthals in Germany. Charles Darwin's work *On the Origin of Species* had come out recently, and a broad section of society was beginning to understand that human beings had been in existence for much longer than the 6,000 years claimed by certain Bible scholars.

Examination revealed that the skeletons from Cro-Magnon were not Neanderthals; they were more like people living today. Europe's first anatomically modern humans were dubbed Cro-Magnons.

Today, the house nearest to the Cro-Magnon shelter is a guesthouse. It is built into the hillside, so some of the corridors have walls of rock. The actual site of the excavations, which now counts among the less impressive ones in Les Eyzies, is just behind the landlady's laundry room.

The modern menu is rather easier on the jaw than the food of the Ice Age, with *café au lait* and buttery croissants for breakfast. The location is just as attractive as when the first people chose to live here. You can still bask in the evening sunlight and gaze out over the river – with a refreshing glass of kir these days.

Ice Age people often chose to live in the mouths of caves or under rock shelters facing south-west, where they were warmed by the sun and the hillside shielded them against the chill north wind. And they nearly always looked out over water.

There is some disagreement among archaeologists about whether the people who lived at Cro-Magnon belonged to the Aurignacian or the Gravettian culture. They lived at around the time of the transition between the two, and the earliest excavations were rather chaotic. However, there are many other sites nearby where work was more systematic, enabling the viewer to trace the whole of prehistory metre by metre and layer by layer.

Abri Pataud, just a few hundred metres from Cro-Magnon, is an example. There are traces of Neanderthals at the lowest level; then comes a level with no remains. The first modern humans, typical representatives of the Aurignacian culture, appeared about 35,000 years ago.

Archaeologists have found bones from six individuals at Abri Pataud: two women, each with a newborn baby; a five-year-old child; and an adult male. The best-preserved skeleton is that of a woman who was in her twenties and about 1.65 metres (5⅖ feet) tall. Her jawbone shows damage resulting from a very serious dental inflammation – so serious that it may have led to a painful death, if it was not childbirth that killed her. Ice Age hunters hardly ever suffered from caries – they consumed too little sugar and starch for that – but wear and tear and inflammation could cause other serious dental conditions.

As yet, no reliable DNA analyses are available from either Cro-Magnon or Abri Pataud. One attempt was made by a

German researcher called Johannes Krause, who started his investigations with the most famous skeleton of all, Cro-Magnon 1, which is kept at the *Musée de l'Homme* (Museum of Humankind) in Paris. Krause attempted to extract DNA from several bones, but only one of his analyses was successful. He then ran isotope tests on the bone in question. By comparing different forms of one of the elements present, nitrogen, he hoped to find out what kind of food the individual concerned had lived on.

As we have seen, the diet of European Ice Age people contained relatively little carbohydrate, but it was rich in protein from meat and fish. Yet the bone that supposedly belonged to Cro-Magnon 1 seemed more like one from a modern vegan – or a cow fed only on grass.

Krause then had the bone radiocarbon-dated, which revealed that it dated back only to the fourteenth century AD. The levels of nitrogen isotopes were entirely plausible for a poor individual living in the Middle Ages on a diet consisting almost exclusively of porridge, with practically no meat. The bone was rapidly removed from the collections at the *Musée de l'Homme*.

The last traces of the Gravettian culture petered out some 20,000 years ago in Les Eyzies, just as elsewhere in Europe. It was followed by a culture known as the Solutrean.

Abri Pataud and various other sites clearly show how the climate cooled down dramatically around this time. In the present day, the average temperature in Europe is about 12 degrees Celsius (54 degrees Fahrenheit). When the Ice Age was at its coldest, between 18,000 and 19,000 years ago, the average temperature was about minus four degrees (25 degrees Fahrenheit). Horses, which had previously been a common prey, became much rarer. The animals that remained were mainly reindeer, bison and a number of predators that were resistant to cold, such as Arctic foxes and wolves.

And humans. Curiously enough, human culture blossomed – as the great museum in Les Eyzies clearly shows.

***

The *Musée National de Préhistoire* in Les Eyzies–de–Tayac is a large, lavishly appointed museum run by the French state. It is partly built into the light-coloured limestone hillside, just like the Cro-Magnon guesthouse.

A whole floor of the museum is given over to Ice Age tools, most of them of stone, though there are also some made of antler, bone and ivory. They are systematically arranged in the display cabinets by chronological period and culture. It is quite difficult for an amateur like me to make out the transitions between Neanderthals and modern humans, and between the Aurignacian and the Gravettian. But the transition from the Gravettian to the Solutrean, which took place around 20,000 years ago, is striking, even for the most inexpert observer.

The tools belonging to the Solutrean culture are dramatically different and far more sophisticated. They are wafer-thin, polished, sharp and aesthetically pleasing. Some of them are so beautifully fashioned and of such exaggerated size that they can hardly have been used for practical purposes; they must have been decorative objects. The flint of which they are made is of a particularly high quality and often came from rocky outcrops 50 kilometres (30 miles) away. These tools were probably made by skilled specialists; knapping flints to fashion these willow leaf-shaped points was not a task that could be performed by just anyone.

On the other hand, it does look as if just about everyone made their own spear-throwers out of antlers. That is apparent from their rather more amateurish shapes and the images incised in them. Spear-throwers, or atlatls, were an innovation that made it easier to hunt in the open landscape of the Ice Age. They enabled hunters to use the principle of leverage and put more power into throwing their spears.

Sewing needles also appeared for the first time in western Europe during the Solutrean period. The museum's display shows how they were fashioned step by step, starting with a mammoth tusk. Only in Russia have older finds emerged, as I mentioned earlier.

People have been wearing clothes for a long time. Mark Stoneking at the Max Planck Institute in Leipzig has used an unconventional method to calculate the age of clothes – analysing DNA from body lice. By comparing different species of body lice, and by making comparisons with head lice and the lice that live on chimpanzees, he has been able to estimate that the history of clothes goes back about 107,000 years. Admittedly, he gives his margin of error as several thousand years, but he is still more precise than any previous researcher has dared to be. Stoneking's analyses of louse DNA also show that people had already begun to wear clothes in Africa.

Archaeologists in many places have found stone scrapers that were probably used to prepare animal skins for making clothes. Even Neanderthals possessed this technology. This skill was a precondition for anyone attempting to live in any region other than the tropics. Clothes must have been absolutely essential, both in cooler parts of Africa and the Middle East, and for the first inhabitants of Europe.

But cloaking oneself in an animal hide and piercing a few holes with an awl so as to join two pieces of hide together, forming a simple tunic, is one thing. Using needles to make anoraks with fur-lined hoods, well-fitting leggings and watertight boots is quite another.

Needles with eyes may not sound particularly impressive to modern-day people, but during the coldest periods of the Ice Age they meant the difference between life and death. Warm, weatherproof clothes made of hide must have been absolutely vital in such a bitterly cold climate, and eyed needles facilitated the task of making them. Needles could also be used to make nets and fish traps. This made it possible to fish and hunt in a more flexible way, enabling all the members of a group to join in, regardless of their physical capabilities. Needles may well be among humankind's most significant inventions.

Clearly there was a technological leap forward in the development of western Europeans here in south-western Europe, at the very point when the Ice Age was at its coldest.

It was Jiří Svoboda, the archaeologist from Brno, who gave me the best explanation for this. He believes that groups of people from more northerly regions of Europe were forced by the cold to move southwards. The various groups met in the new, cold and challenging environment and pooled their knowledge. This conglomeration of people with different skills became an ideal breeding ground for development and innovation.

Genetic research supports the hypothesis that people migrated from northern Europe to warmer places of refuge during the coldest period of the Ice Age, between 25,000 and 18,000 years ago. The locations where they took refuge were in various regions of southern Europe, such as the area around the Black Sea, present-day Greece and Italy, and further east in Siberia.

I have reason to believe that people related to me in the direct maternal line spent the coldest years of the Ice Age here, near Les Eyzies-de-Tayac in south-west France, or in northern Spain.

The results from the Icelandic company deCODE Genetics told me that I, in common with about 1 in every 10 Europeans,

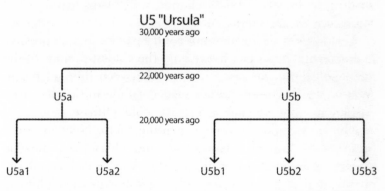

Source: Malyarchuk 2010

Figure 3 **Ursula's daughters.** *The woman called U5 or 'Ursula' lived in Ice Age Europe over 30,000 years ago. The branch labelled U5b1 probably originated in Spain or south-west France during the coldest phase of the Ice Age.*

belong to the U5 group. But in the summer of 2011 I got in touch with some Swedish family history researchers who had started to take an interest in the opportunities offered by DNA analysis. We took a closer look at the results I had received from deCODE and worked out that I belong to one of the two subgroups of U5, namely the one called U5b. That, in its turn, is divided into three smaller subgroups, and I belong to the first of these, U5b1.

There is much to suggest that U5b1 emerged within the group that sought refuge in south-west Europe during the coldest period of the Ice Age; that is, when the region was dominated by the culture we know as the Solutrean. One important clue is where the most variation exists among present-day people. In the case of U5b1, variation appears to be greatest in the regions around the Pyrenees, in south-western France and northern Spain.

Another quite unexpected piece of evidence was published in 2005. In this, a group of Italian researchers showed that a subgroup of U5b1 found in nearly half of all Sami people is closely related to one found among the Berbers, the original inhabitants of North Africa. Their lineages seem to have diverged a few thousand years ago. This new information was startling, as there are 4,000 kilometres (2,500 miles) between North Africa and northern Scandinavia. The most credible explanation is that people belonging to the U5b1 group migrated in different directions; some went northwards when the Ice Age loosened its grip somewhat, so that some of their descendants ended up in northern Scandinavia. Others went southwards, crossing the Strait of Gibraltar and migrating into Africa.

A third line in the evidence is provided by DNA analyses of fossilised human bones from western Europe. At the moment there are only a few such analyses of sufficiently high quality, and they are from a period later than the Solutrean. However, these analyses lend some support to the idea that U5b1 was common in south-west Europe during the coldest phase of the Ice Age.

One of the very best sources of finds from the Solutrean culture is called Laugerie Haute. It took me about half an hour to walk there from the Cro-Magnon guesthouse. Laugerie Haute is, in fact, one of the very best sources of finds from all the cultures that have existed in the region over at least the last 20,000 years. Archaeologists have excavated 43 layers, ranging from the Aurignacian culture to the Solutrean and the Magdalenian culture that succeeded it.

The huge rock shelter seems to have provided a shared camping area, a spot where several smallish groups congregated for part of the year, especially in the autumn, when they had best access to their prey. About 100 people seem to have lived here at any one time, corresponding to three, four or five smaller groups. There was no need for them to live under cramped conditions; the sheltered area under the rock overhang is the size of two tennis courts. Most of the roof has fallen in now. The ground is strewn with enormous rocks.

Like so many other Ice Age dwelling places, Laugerie Haute is just a few metres away from a river. The people who stayed here enjoyed evening sunshine and a waterside view – as in the most desirable residences on the books of a modern estate agent.

One can try to imagine how the people of the Solutrean felt when they gathered here in the depths of the Ice Age. I fancy it must have been a pleasure to come here. The various small groups probably lived in isolation for much of the year, hemmed in by the bitter cold. At Laugerie Haute they could meet other people, celebrate and hold ceremonies. The young had an opportunity to meet a partner, people could sit by the fireside sharing experiences and everyone had enough to eat.

Their diet consisted chiefly of meat and bone marrow from reindeer. The numbers of other prey, such as horses, fell dramatically in the region when the cold was most severe. Ice Age people did eat plants, as evidenced by microscopic traces on their teeth. But the growing season was short, especially during the cold Solutrean.

With all due respect to excavations and stone tools, it is the cave paintings that are the main draw for the hundreds of thousands of tourists who visit Les Eyzies every year. I, too, set off on an art tour of the region's caves.

***

Laugerie Haute feels like visiting the remains of a huge party venue where my forebears used to hold their biggest celebrations. The Cap Blanc rock shelter (Abri du Cap Blanc), which lies about six kilometres (3¾ miles) east of Les Eyzies, has a more intimate feel to it. Coming here is more like visiting an individual family and admiring their interior decor.

Just six of us have gathered together here at the small museum, with our pre-booked entrance tickets. There is no room for any more visitors. We will be allowed in to look at the site for a few minutes. Having checked carefully that none of us have smuggled a camera in, the guide unlocks a heavy iron door.

Archaeologists have argued about when the friezes running along the rock walls were actually created; nowadays they are assigned to the Solutrean period. They are unusually well-preserved examples of how Ice Age people decorated the places where they lived from day to day. At most other sites, the rock walls that once bore these decorations have been eroded away.

The Cap Blanc friezes are reliefs chiselled directly into the rock wall, forming a long parade of horses and bison. Oddly enough, these animals were not particularly common in people's diet. That can be seen from the animal bones that archaeologists have discovered at the site. Almost 95 per cent of them come from reindeer, which were people's most common prey during the coldest periods. Horses and bison clearly had great emotional significance. The artists of the Ice Age often chose to depict quite different animals from those they hunted day to day.

From the homelike setting of Cap Blanc, I make my way down the valley to Font-de-Gaume. I am about to experience something entirely different that, far from reflecting everyday life, verges on the spiritual. This is art of the most exclusive kind, hidden deep within a cave. The Ice Age artists who created these images must have been the Rembrandts and Leonardos of their time.

I go in with a French-speaking group of 10. Just a few steps into the mountainside all sensory impressions from the outside world cease. The light of the sun is extinguished. There is no birdsong, no breeze to be heard now. My skin registers only cold and damp. A low humming sound arises in my head, as my hearing tries to compensate for the silence within the mountain. I blink as my eyes attempt to adjust to the dark.

Although the passageways in Font-de-Gaume are narrow, at least we modern visitors can walk upright, as the floor has been lowered for the sake of our comfort. Moreover, we have electric torches, which the guide carefully switches off as soon as we leave each section of the cave. He is anxious to protect the pictures as much as possible, both from electric light and against the acid produced by visitors' breath.

The Ice Age artists were obliged to crawl through certain sections. But those who painted the images in this cave had access to an important new invention – lamps. That meant they were no longer restricted to using simple wooden torches, as previous generations had been. The artists in Font-de-Gaume lit their way using stone lamps with small hollows, in which they burned animal fat and wicks made of vegetable matter.

Font-de-Gaume, located on the periphery of Les Eyzies, was in use some 17,000 years ago, at the time of the Magdalenian culture. A continuation of the Solutrean culture, the Magdalenian was an undramatic transitional phase. While tools became slightly more sophisticated, the people seem to have been essentially the same. The cold of the Ice Age was gradually beginning to loosen its harshest grip, and the climate was becoming slightly milder.

The artists of Font-de-Gaume painted their pictures with mixed pigments: yellow, red, brown, black and many intermediate shades. They produced these colours from reddish iron oxides and black manganese oxide from the mountains nearby. Sometimes they would burn stones to create particular colours. The artists ground the pigments in mortars and mixed colours on palettes, just like their modern counterparts. They applied the paint to the cave walls in many different ways, using their fingers, sticks, brushes made of animal hair, or feathers. They also blew powdered pigment directly onto a damp surface, just like the fresco painters of the Renaissance.

Bison, with over 80 individual animals depicted, are the most frequent motif at Font-de-Gaume. Each is painted in such a way as to reveal its individual characteristics. The artists knew the sex of the animal they were painting, its age, the season and the situation in which the animal found itself. With skill and foresight, they exploited the shapes of the rock wall to achieve three-dimensional effects. In one of the chambers, some of the paintings are five metres (16½ feet) above the floor of the cave. To reach that high, the artists must have stood on each others' shoulders.

Polychrome paintings can be found in only a few Ice Age caves. The best known are at Altamira (Spain); Lascaux, near Les Eyzies; and Chauvet, in the mountains west of Lyon. The public is no longer allowed into these three caves at all. The only option is to look at copies.

At the time of writing, members of the public can still enter Font-de-Gaume and view the original polychrome cave paintings, provided that they queue up on the spot and buy tickets the same morning. Whether Font-de-Gaume should also be closed, to protect the paintings against the depredations of lighting and visitors' breath, is under discussion.

I am grateful that I came in time to wander around in the cool, damp darkness and see bison, mammoths and horses appearing in the dim light, a little worn after all these thousands of years, but still so skilfully executed that they almost come alive.

The pictures from a few thousand years later are simpler, still skilfully executed, but more stylised. In the cave at Les Combarelles, a few kilometres away from Font-de-Gaume,

horses predominate. There are drawings of cave lions, cave
bears and mammoths as well, but horses are the most common
motif. In some cases, their shapes are simply incised into the
rock wall, in others they are filled in with black pigment
made from manganese oxide.

The guide who escorts a French family and me through
the long, narrow passageways believes that Les Combarelles
was used by a clan whose totem animal was the horse. They
returned to the same cave again and again over a few
thousand years and drew hundreds of pictures. This was a
way for the clan's shaman to gather spiritual powers, or so
he believes.

To reach the cave at Rouffignac, I have to hire a car and
drive about half an hour from Les Eyzies through a landscape
of hillsides covered in deciduous woods and vineyards. The
pictures here also date back to the Magdalenian and are
approximately 15,000 years old. They recall the incised,
monochrome horses at Les Combarelles.

But it is images of mammoths that predominate at
Rouffignac. There are over 150 of them. The first one meets
you right at the entrance – a cheerful, rotund little woolly
mammoth drawn only about a metre (3¼ feet) above ground
level. It might well have been drawn specially for children.
Further into the cave you can also see bison, horses, mountain
goats, woolly rhinoceroses and a cave bear. At an inaccessible
spot, far down in a hollow, there is even a picture of a human
being, a roughly sketched head in semi-profile.

A total of nearly 10 kilometres (6¼ miles) of tunnels
meander within the mountain. We visitors trundle round on
a little electric train. The lighting comes on automatically at
each stopping point, just for the few minutes when the train
comes to a halt. This minimises our impact on the pictures.

The cave, which is located on a large farming estate, is
managed by the family that owns the estate. Locals have
known of it for several hundred years, and generation
after generation has visited it to scrawl their names on the
walls. But systematic archaeological research began only in
the 1950s.

Frédéric Plassard, the son of the present family, grew up near the cave. He now works here full-time, but has also earned a doctorate in archaeology at the University of Bordeaux. We sit talking for a long time on a bench in the mouth of the cave, in the cool shade of the oak trees, and I try to understand what impelled people 15,000 years ago to make their way into the darkness of the cave to produce these works of art.

There's nothing particularly odd about the fact that they went into the cave, says Plassard. That can be put down to normal human curiosity. People have always wandered around in the cave; we can see that from the eighteenth and nineteenth century scribblings. The Magdalenians had lamps just as we do, although they were simple stone lamps that burned animal fat, and it only takes about half an hour to walk a kilometre (⅔ mile) into the mountainside.

The remarkable thing is the art. It is so well executed that it must have been produced by highly specialised artists. Plassard thinks there were only three or four of them. They may have produced all the works of art on a single occasion, in the course of a few hectic hours. Their work didn't necessarily have to be seen. Creating the pictures was more important than viewing them.

The European cave paintings represented something very special for the people of the time, of that Plassard is convinced. He points out that there are only about 20,000 known cave paintings in the whole of Europe, although the people of the Ice Age lived here for 30,000 years – fewer than one a year for the entire European population of several thousand people. By comparison, the Aboriginal people of Australia have produced millions of rock paintings.

There is one myth that Plassard would like to scotch straight away. Archaeologists used to think that Ice Age artists grew increasingly skilful over the millennia. The animals at Font-de-Gaume, for instance, with their perfect proportions and perspective, are more sophisticated than the clumsier horses in the cave at Pech Merle, which is a few thousand years older. However, the 1990s saw the discovery of the cave at Chauvet, further east in southern France. The art there is peerless – a multiplicity of skilfully depicted polychrome animals. The painting of these pictures may have started 32,000 years ago, as far back as the time of the Aurignacian culture.

This means that we humans have had full potential to produce figurative art for at least that long. Nor should we forget the ivory flutes and tiny statuettes from the Swabian Jura, the oldest being over 40,000 years old. There have been masters, as well as less high-calibre artists, for as long as humans have inhabited Europe.

Sitting in the mouth of the cave, we also discuss theories advanced by the French archaeologist Jean Clottes and his South African colleague David Lewis-Williams. This duo seem to be the main ideologues represented in the region around Les Eyzies. Jean Clottes's books, above all – both the popular science books and the more academic works – are on sale everywhere in shops and museum gift shops. Their theories can be summarised just as the guide in the horse cave of Les Combarelles put it: creating pictures in caves was a way for shamans to acquire spiritual powers.

Frédéric Plassard agrees that the shaman hypothesis is reasonable, though very hard to prove. He urges caution. There may have been different kinds of motive in the course of a period lasting over 30,000 years, in an area stretching from the Atlantic seaboard to Siberia.

Jean Clottes and David Lewis-Williams deploy three lines of evidence in their arguments. The first of these is their own profound knowledge of such images. That is uncontested. Clottes, for example, was one of the leaders of the research carried out in the Chauvet cave.

The second part of their argument is more or less accepted by most archaeologists and anthropologists. Clottes and Lewis-Williams draw parallels between the European artists of the Ice Age and traditional hunter-gatherer peoples in the nineteenth and twentieth centuries. Their books draw on sources including descriptions of Siberian nomads in the anthropological literature, and their own visits to Native

Americans in California and the San people (Bushmen) of South Africa. One of the most important documents comprises over 12,000 pages of interviews of San people, from the end of the nineteenth century to a few decades later by the German linguist Wilhelm Bleek, his sister-in-law Lucy Lloyd and Bleek's daughter Dorothea.

Many of the parallels described by Clottes and Lewis-Williams are undeniably striking. In the shamanistic world view, the universe is often divided into three layers: earth, the domain of us ordinary mortals; the heavens; and the underworld and/or the dark waters of the underworld. Certain animals have the power to reach the heavens or the underworld. Waterfowl, for instance, have access to all three levels, as they can fly and swim, and snakes can slither down into the underworld. Such animals can help shamans to travel to the spirit world.

The caves could represent a way of approaching the underworld, and thereby the spirits and the dead. Lewis-Williams writes about how the rock wall in the cave could act as a membrane – indeed, as the very border between the human and the spirit world. He believes that the artists literally saw the images of animals projected on the walls. All they really needed to do was fill in their internal images. Maybe they had worked themselves up into a trance through lack of sleep, drugs, rhythmic music and frenetic dancing, through the high levels of carbon dioxide in some of the caves, or simply through the sensory deprivation that affects anyone who spends a significant amount of time alone in a dark cave. Although they may not have been in a trance while they were actually painting the pictures, he believes they were reproducing images that had appeared to them in a trance.

The third aspect of the evidence presented by Clottes and Lewis-Williams is the most controversial. They believe that the first art originated not only in the trance of the shamans, but – above all – in the way the human brain typically functions when in a state of hallucination or psychosis. For example, they interpret the zigzag patterns found on 70,000-year-old stones in the South African cave of Blombos as representing light phenomena that appear to many people suffering from an attack of migraine. According to Clottes and Lewis-Williams, these phenomena are a kind of mild hallucination. In their view, the mammoths, horses and bison found in cave art represent more profound hallucinations.

Personally, I am sceptical about whether these cave paintings represent hallucinations. In my view, simple human creativity is quite sufficient as an explanation, possibly intensified by a trance-like state.

CHAPTER SEVEN

The First Dog

Some of my relatives remained in the region around Spain and south-western France, while others with our shared maternal lineage migrated southwards. They crossed the Strait of Gibraltar into North Africa, moving along the coastline, into the mountains of Kabylia and as far southward as Senegal. Their descendants can be found there to this day. We know this thanks to the presence of mitochondria belonging to the U5b1 haplogroup in these regions. They are rare, admittedly, but they do exist.

Others began to migrate northwards when the worst of the cold was over. They followed the reindeer that were their main prey. Some of them went directly northwards along the river that we today know as the Rhine.

About 14,700 years ago, the climate in northern Europe began to grow a great deal warmer. The steppe filled with trees such as birch, willow and aspen. Deciduous woodland of this kind is not a suitable habitat for reindeer, as it provides nothing for them to eat in winter. My forebears in the Rhine Valley had to adjust rapidly to hunting other animals, such as elk (moose), deer and beaver – a major change.

Two of my relatives died near today's Bonn: a man in his fifties and a woman of about 20. Their companions dug a grave, laid them in it side by side and scattered copious amounts of red pigment over them. As grave goods, the dead were given a fine hairpin made of bone, a patterned piece of antler and a deer's tooth painted red. And they were accompanied to the realm of the dead by a dog. This must have represented a major offering, the greatest gift they could have been given.

The burial find, known as Bonn-Oberkassel, has been dated as being about 14,500 years old, from just after the end of the Ice Age Magdalenian culture. DNA analysis of the man's and the woman's mitochondria shows that both belonged to the U5b1 haplogroup through their maternal lines. That is the group I belong to, which means that the two people in the grave and I had a common foremother who lived a few thousand years earlier. They were the children and grandchildren of 'Ursula', just as I am.

But the really significant point about the Bonn-Oberkassel grave is not that my distant relatives lie buried there. The most important find is the dog's skull. This is the most ancient dog on which there is a scientific consensus. It meets all the possible criteria that anyone can establish for an early dog. Its appearance is right, the time is right, the place is right and the dog's DNA set is right.

***

The question of when wolves became dogs has been the subject of heated debate for many decades. There is a great deal of prestige at stake. Dogs were our very first domesticated animals and are still known as 'man's best friend'.

In the past, archaeologists tried to distinguish dogs from wolves solely according to the appearance of their fossilised bones. Their main assumption was that dogs which had benefited from the care of humans for many generations could be expected to become smaller and slighter than wolves, and to look different in some way.

On the basis of these definitions, there are many possible early dogs, the oldest of which is over 30,000 years old. There are fossil remains from Russia, Ukraine, the Czech Republic, Switzerland and Belgium. Some of the finds that are claimed to be early dogs come from the excavations at Dolní Věstonice in the Czech Republic, described in Chapter 5.

Over the last 20 years, geneticists have joined the discussion, basing their arguments on DNA analyses of varying degrees of exhaustiveness and calculations that vary in sophistication. They have still not reached a consensus, to put it mildly. However, the current state of knowledge can be summarised as follows.

The ancestor of today's dogs appears to have been domesticated at least 15,000 years ago, and possibly much earlier. The place of birth of the first dog was probably somewhere in Europe or Siberia, though China cannot be ruled out.

People may well have tried to tame wolves even earlier. There are 30,000-year-old fossils from the Altai Mountains of Siberia and Belgium's Goyet cave that resemble dogs, which may well indicate that such attempts were made. However, they do not appear to be the forerunners of modern dogs, but genetic dead ends.

All of today's dogs – African basenji, the wild dingoes of Australia, blue-eyed Siberian huskies, neatly blow-dried miniature poodles and playful Labradors – seem to have a common origin in one specific group of wolves from Europe or Asia. These wolves are probably long extinct, which makes it more difficult to establish kinship. But when researchers compare DNA from fossil wolves tens of thousands of years old with both fossil dogs and contemporary specimens, the pattern emerges more clearly.

One of the problems facing researchers is that domesticated dogs have sometimes interbred with wild wolves, as shown by the DNA of hunting dogs in Scandinavia and half-wild dogs in China, for example. This interbreeding makes the picture fuzzier. However, the basic pattern remains. When we left Africa, we had no dogs with us. It was not until we reached Europe and Asia that we came across wolves. We tamed them at some point during the Ice Age, during a period when we were still hunter-gatherers, and before some of us crossed Beringia and made our way down into the Americas.

The question is why. What use were dogs to us?

***

Over the years, I have interviewed nearly a dozen of the world's leading dog researchers, and have been given nearly as many diverging explanations for why dogs and people started to live together.

Many researchers believe that it was not people that domesticated dogs, at least not initially – it was dogs that domesticated us. Ice Age people hunted wolves for their pelts, and wolves must have regarded us as dangerous. But they also saw the advantages that humans offered. We left large piles of food behind us, the remnants of our prey that we were unable to use ourselves. We dumped the carrion left over from our hunting on the edge of our temporary settlements, as it stank and attracted predators.

Wolves would go in search of these leftovers at night, when people had gathered round their fires or were asleep. Sometimes, especially at dawn and dusk, humans and wolves might encounter one another. That could end with the human killing the wolf. But on one occasion the wolf was simply too appealing. Maybe it was a little cub, a good-natured trusting little pup that no one with a heart could bring themselves to kill. So instead, the little wolf cub was allowed to sit at the fire among the humans, and to play with the children. Anyone who has ever seen a young child and a puppy playing together will know just what I mean.

The millennia passed, and wolves that were able to control their aggression and endear themselves to people had found a new niche enabling them to survive. This was a case of what biologists call 'artificial selection'; traits valued by humans were more likely to be passed on to the next generation.

Many scientists have examined which genes differentiate wolves from dogs. A large proportion of these genes affect the brain first and foremost. The US researcher Robert Wayne, for example, has detected a particular genetic mutation that seems to be present in all dogs, but not in wolves. A similar mutation can be found in people with the congenital condition known as Williams–Beuren syndrome. Such people generally have slight learning difficulties. But the most striking feature of people with Williams–Beuren syndrome is their friendly, outgoing and trusting nature.

Many other features that differentiate dogs from wolves – in terms of both character traits and appearance – have to do with the more childlike nature of dogs. They resemble wolf cubs more than adult wolves; they are playful and good-natured rather than serious and intense. In many cases, dogs also have more stubby noses and shorter legs than adult wolves, just like wolf cubs.

Moreover, dogs also have an exceptional ability to read people's thoughts. Many experiments have shown how they can understand what we want. They can follow our gaze and look at the things we point out. Other animals, such as chimpanzees, wolves and cats, can be at least as intelligent in many other respects. Yet they cannot compete with dogs in tests based on understanding human behaviour.

The nights were cold during the Ice Age, and people may have used their first dogs as cushions, to keep warm. The archaeologist Lars Larsson told me that even today Australian aboriginals speak of 'one-dog', 'two-dog' and 'three-dog nights', the 'three-dog night' being the coldest of all.

Others believe that people first benefited from wolves' function as guard dogs. They would lie at the edge of the camp, feasting on leftover meat. Little by little they began to spend the night there. Wolves and dogs sleep far more lightly than humans. If another, more dangerous predator approached – a lion, for instance – they would begin to howl loudly. This would wake the humans, who could then defend themselves.

Helping humans to hunt must have been one of the functions dogs fulfilled from an early stage. One of wolves' inherited behavioural characteristics is the propensity to hunt in packs. Much has been written about how hunting dogs may have enabled humans to hunt more effectively, especially the big game of the Ice Age, such as mammoths and woolly rhinoceroses. According to some theories, the large animals in Europe, Asia and the Americas faced rapid extinction once humans and dogs began to hunt together. But there is no consensus on this. Some new findings indicate that dogs and humans are blameless in this regard; mammoths and woolly rhinos are more likely to have died out because the climate became warmer, bringing changes in vegetation.

Martin Street is the archaeologist who has done most work in recent years on the dog from Bonn–Oberkassel. His theory is that what is known as 'putting the game at bay' was one of the first important tasks performed by dogs. This is a method of hunting still used today in many places, including the forests of Sweden. The dog runs around in the woods on its own to track game, while the hunter tries to stay near it. Once the dog locates its quarry, it starts to bark, forcing the animal to stop moving and focus on the dog's irritating barking. The dog has put its quarry at bay. In the meantime, the hunter creeps nearer and shoots the animal.

This type of hunting emerged when woods started to grow on the tundra, blocking the view. Before that time it was easier for hunters to scan the landscape for their prey from an elevated point. This is what makes it so interesting that the first dog universally recognised as such, the one from Bonn–Oberkassel, lived 14,500 years ago, at precisely the time when the tundra of the Ice Age was beginning to give way to woodland. That circumstance, in my view, is rather too striking to be a mere coincidence.

If we began to use dogs even earlier, during the colder periods of the Ice Age, I believe transport must have been one of their major functions. Dogs may have been used as pack animals and may also have pulled sleds and people on skis. They may have enabled us to move around and maintain networks over very considerable distances. Admittedly, no devices that would have enabled dogs to carry a load or pull sleds or skis have been preserved. However, such artefacts would have been made of wood and other organic material, and the chance of such substances surviving tens of thousands of years is minimal.

Conceivably, the first function dogs fulfilled for people may have been to provide a source of food. If this was the case, keeping domesticated dogs would have been a way of ensuring access to meat in hard times. The researcher who mentioned this possibility to me was Peter Savolainen, one of the foremost advocates of the theory that dogs originated in China.

Savolainen, who works at Stockholm's Royal Institute of Technology, was among the first researchers to conduct a large-scale comparison of dogs' mitochondrial DNA; I reported on his study in *Dagens Nyheter* back in 1997. He and his collaborators attended dog shows and collected hair from several hundred dogs, which they then compared with wolves from different parts of the world. The original purpose of the research was to enable forensic technicians to use DNA profiles to identify the breed of a dog that had left traces at a crime scene. However, the researchers soon realised that their work could also be used to identify dogs' place of origin. Their results suggested that wolves first became dogs in South-east Asia, which happens to be a region where people eat dog meat to this day.

When in Hanoi, northern Vietnam, which lies within the region identified by Savolainen, I once visited a curious street full of specialist restaurants – all cheek by jowl – whose menus featured dog meat. Some Vietnamese people like taking Europeans there to tease them about the habit of eating dog meat, as they know full well that we generally regard it as totally taboo. The Vietnamese journalist who accompanied me in Hanoi explained that he and the people he knew kept dogs as pets, just as Europeans do, and developed strong emotional ties with them. Yet they may also eat dog meat on occasion. There are, quite simply, two contrasting attitudes.

A similar pattern is also discernable in excavations from the European Stone Age. Some dogs were clearly highly respected family members honoured by distinguished burials. Other dog bones bear scratch marks from tools – which may well indicate that people ate their flesh. Both types of remains have been found at Lake Hornborga in Sweden, the oldest being about 10,000 years old.

Peter Savolainen's early conclusion that dogs originated in South-east Asia is strongly contested today, particularly since other researchers have begun to analyse DNA from ancient fossils of dogs and wolves. Today there is at least as much evidence to suggest that dogs originated in Europe.

Savolainen may, of course, be correct in his theory that dogs' first role in relation to humans was to provide a reserve of food. But they may also have been companions and playmates, warm cushions, a means of transport, guard dogs and hunting dogs. One possibility need not exclude the other. Beyond a doubt, our first function as far as dogs were concerned was to keep them fed.

Nor should we underestimate the importance of affection. Surely the feelings that today's dog owners have towards their animals also existed during the Ice Age. After all, those of my kinsfolk who lived in the Bonn-Oberkassel region 14,500 years ago – the man and woman from the U5b1 haplogroup – were accompanied by a dog on their way to the happy hunting grounds.

CHAPTER EIGHT

Doggerland

The dog, the man and the woman from the Bonn-Oberkassel burial site lived at a time of great upheavals. Just a few centuries earlier, parts of Europe had felt the impact of a pulse of warmth that augured the end of the Ice Age.

The harshest period of cold was over. The earth had been slowly thawing out for several millennia. The path of our orbit around the sun had changed, following the repeated cyclical phases known as Milankovitch cycles. More and more solar energy was reaching the earth. Melting snow and ice left dark water and soil exposed, further heightening the earth's capacity to absorb more solar radiation.

In addition to this global warming, temperatures changed dramatically at a regional level. Within just a few hundred years – possibly even faster – the average temperature in north-western Europe rose by several degrees. This was probably due to shifts in Atlantic currents.

It may sound as if the growing mildness of the climate, once so cold, must have been a blessing for Ice Age people. Yet I am not sure they saw it that way to begin with. The change seems to have been so rapid that they had no time to acclimatise. Within a mere generation or so, the hunters of the Ice Age were forced to change a way of life that had served their forebears well for many thousands of years.

In the longer term, the warmer climate brought a great improvement in living conditions. Higher temperatures and more precipitation meant richer vegetation and hence more prey. This enabled more people to survive and reproduce. DNA analyses show that the hunters who peopled Europe multiplied rapidly at the end of the Ice Age. But now they had to learn to hunt the new kinds of prey that thrived in the new forests, or to follow the reindeer on their migrations north and eastward.

Those who followed the reindeer due north from Bonn-Oberkassel would have reached Doggerland, a land mass that no longer exists. Today it lies beneath the North Sea. Sometimes the Dogger Bank is mentioned in weather reports. But Doggerland, now part of the seabed, once extended from Denmark to Scotland. When the land mass was at its largest – about 20,000 years ago, during the coldest period of the Ice Age – it probably stretched as far north as the Shetland Islands. Doggerland was separated from the Norwegian coast by a narrow strip of deep sea, corresponding to what we now know as the 'Norwegian trench'. Midway between the Shetlands and the area now occupied by the Norwegian city of Bergen were mountains, dubbed the 'Viking-Bergen Island' in retrospect.

For certain periods, Doggerland may well have been one of the most favourable habitats anywhere in the Europe of the time: fertile land criss-crossed by freshwater rivers, with ample access to game.

In a display cabinet in the Danish National Museum, Copenhagen, I have seen tools and art objects made of bone and antler that were fashioned many thousands of years ago by people living in Doggerland. Some of them have turned up in fishermen's nets, while others were found washed up on Danish beaches. Underwater archaeologists have searched for sunken settlements off the coasts.

Back in the nineteenth century, oyster fishermen began to find strange bones from mammoths and reindeer in the waters off the coast of Britain. In 1931 the British trawler *Colinda* dragged up a barbed spearhead made of antler that has been dated at nearly 12,000 years old. Since then, fishermen, divers, archaeologists and geologists have discovered many objects that tell us how prehistoric people lived in the area that is now part of the seabed. But the latest information about the sunken land comes from a different and unexpected source.

***

I travel to Bradford, England, to interview the archaeologist Vincent Gaffney, the director of a major project that resulted in – among other things – *Europe's Lost World: The Rediscovery of Doggerland*. This book, published in 2009, is the most extensive overview of the subject to appear so far.

Had I been travelling 10,000 years ago, I could have walked dry-shod all the way from Sweden to Bradford. Instead, I fly to London and take the train to northern England. Despite the great beauty of the green, rolling countryside that surrounds it, the inner city itself strikes me as depressing. The glory days of the textiles industry are long gone. Few visitors come here of their own accord these days. A woman on the train asks me why on earth I want to visit Bradford.

But the University of Bradford clearly has a strong reputation. It turns out that Vincent Gaffney was appointed there after a conflict with the University of Birmingham, his previous employer.

We are meant to be meeting in the morning, but at the last minute Gaffney is called away to a meeting at his new university. Instead, we meet late in the evening in the hotel bar. It is an unusually muddled interview. I confine myself to two small ciders, but Gaffney manages to down three large beers. A red-haired man in his sixties, he is already pretty lively at the start of the evening. As the hours pass and the level of beer in his glass sinks, I find it increasingly difficult to channel his verbal outpourings. However, the flashes of wit and the wild associations he comes out with are underpinned by brilliant, groundbreaking archaeology.

Gaffney began his career as a specialist in Roman remains in the Mediterranean region, but over the years he came to focus on remote analysis – a variety of methods for investigating land at a distance, without the need for excavation. He was running a course on the subject for doctoral students at the University of Birmingham when a student asked which region he most longed to research. 'Doggerland' was the obvious answer, as much of European prehistory is likely to be preserved there, on the seabed. The young doctoral student then proposed they work together, using data from oil companies prospecting for gas and oil. That was an option Gaffney had not previously considered.

They managed to link up with a geologist with expertise in surveying North Sea oil and gas deposits – exactly the relevant field. With his assistance, they were able to access a large quantity of data, which they processed by computer.

The technology they applied, 3-D seismic surveying, normally serves to probe layers much further down in the seabed, where oil deposits can sometimes be found. The archaeologists, however, wanted to conduct a more superficial survey, going down just a few metres. They were delighted that the method proved equally effective for their purposes. Within just 18 months, the research team was able to put together a detailed map of a section of Doggerland the size of the Netherlands. They created an image of a whole landscape, with lakes, wetlands, estuaries, mountains and plains. It was crossed by a large river, which they named the Shotton, after a distinguished geologist.

Unfortunately, there is one question that appears difficult to answer. What did the periphery of Doggerland look like? How far out into the Atlantic did the land mass extend at different times? I repeat my question several times in different ways, but finally Vincent Gaffney says I'll have to accept that they don't really know.

The best maps currently available were drawn up in the late 1990s by the British archaeologist Bryony Coles, from Exeter. Her work was based mainly on modern measurements of the relief of the seabed. At that time, researchers had just established that the sea level had risen about 120 metres (390 feet) since the coldest period of the Ice Age, 20,000 years ago. It was therefore reasonable to assume that all the areas of the seabed lying less than 120 metres (390 feet) down must have been dry land at one time or another. That would imply that areas which are much shallower, such as the Dogger Bank, would have been high mountains. However, the match is not quite that simple. The major rivers that once flowed across Doggerland bore sediment that accumulated over the millennia. The landscape changed both in surface area and relief. Bryony Coles's maps are thus based on reasonable assumptions, rather than on hard facts.

Analyses of pollen from cores sampled from regions bordering on Doggerland reveal how the vegetation changed. The predominant landscape type during the coldest periods of the Ice Age was dry tundra. Unlike the interior of Scandinavia, Doggerland was probably not covered by thick ice sheets. Gaffney believes that small groups of Ice Age hunters visited the region even during the coldest periods of the Ice Age. One indication of this is the flint tip of a weapon, discovered by pure chance in the course of drilling in the seabed off the Viking-Bergen banks.

At the end of the Ice Age, the sea level rose and Doggerland shrank. Yet the remaining areas of dry land became all the more fertile. Woods appeared, with small birches and willows first, gradually giving way to pines and hazel; hardwood broadleaved trees such as elms, lime trees and oaks came later.

Gaffney hopes to secure grants that will enable him to pursue his research. He wants to have drilling carried out on a large scale in those areas of Doggerland that he thinks most likely to have been populated. These are mostly around estuaries and river mouths, along rivers and, above all, around the great 'Silver Pit' lake.

He has already had drilling carried out in one or two places. He has subjected deposits from a settlement identified on the seabed off the Isle of Wight to a new type of DNA analysis. The whole core sample is analysed together at the same time, then researchers use computer-aided analysis to try and identify the organisms of which traces have been found in the sample. This produces detailed information about the diet of the people living in the settlement. Different species of plants and animals can be identified from the DNA sample.

Gaffney is convinced that Doggerland was one of the heartlands of the hunting cultures that peopled north-western Europe after the end of the Ice Age. It afforded them the best hunting and fishing. Present-day England and Scotland, on the other hand, were bare, inaccessible mountainous regions at that time. They had little to offer. The inhabitants of Doggerland may have migrated there for relatively short periods in order to hunt.

The people of the time probably also regarded the southern part of today's Sweden as temporary hunting grounds. Large tracts of Sweden were still covered by ice 14,000 years ago. The southernmost tip and the west coast of Sweden, however, were ice-free and received occasional visitors. To reach Sweden, all people had to do was cross a few rivers. There were periods when they could travel all the way from Doggerland to the region now called Skåne, which forms the tip of southern Sweden – without crossing any bodies of water.

**

The very first people to visit Sweden were probably a group of young adventurers living about 14,000 years ago. They belonged to what is known as the Hamburg culture, after the north German city of that name. This was a culture of reindeer hunters who followed the migrations of their quarry over vast areas. Doggerland was one of the heartlands, but the same culture extended from the region that is now Belgium in the west, through Denmark and northern Germany, to Poland in the east.

To travel from Denmark to Sweden, the adventurous young reindeer hunters had to cross a broad and very rapid torrent. It must have been a hazardous undertaking to get across by boat, particularly at the narrowest point, between Helsingør on the Danish side and Helsingborg in Sweden. To the east was the Baltic ice lake, a great lake formed from meltwater from the ice sheet. As it was much higher than the North Sea, water poured out through the Öresund Strait with great force. Presumably it would have been safer for these young people to paddle a little further north and land in the region that is now the Swedish county of Halland.

They may have crossed the ice in winter. But that was also dangerous, as the ice sometimes broke. They would have had to tap it with their spears at very regular intervals to check its thickness and stability.

Yet it may have been worth their while to take up all these challenges, as reindeer and horses thrived on the margins of the great glacier in summer. The same applied to the best fur-bearing animals, wolverines and Arctic foxes. And today's archaeologists believe an expedition to new, unknown land would have conferred honour and prestige. The young adventurers returned to the new land again and again to hunt, but never stayed very long.

Thousands of years later, when the land beyond the Öresund Strait had already been known as 'Sweden' for several centuries, collectors would sometimes stumble across a flint arrowhead left by the first visitors.

For a brief period – 300 years, perhaps – there was probably a land bridge between Denmark and Sweden. By that time the climate had again become somewhat milder. The reindeer-hunting Hamburg culture from the tundra was replaced by the Bromme culture, which was more specialised in hunting deer, elk and other animals suited to life in birch woods. Everything suggests that this was essentially the same group of people; it was just that they adapted their culture somewhat as the climate grew warmer, woodlands emerged and their prey changed. The waters of the Öresund Strait seem to have dried up about 13,000 years ago, after which reindeer, deer, bears, beavers and people streamed over into Sweden. The people of the Bromme culture were active in Skåne for several generations.

Then the Öresund Strait was flooded once again. The Atlantic currents changed course again, and the cold returned. The Ice Age reappeared in north-western Europe, as what geologists call the Younger Dryas. The birch woods disappeared, replaced by tundra, lichen and flora such as mountain avens (*Dryas octopetala*). This period of cold lasted for over 1,000 years. For most of this time, Sweden was so cold that it would not have been habitable. It was not until the end of the Younger Dryas, when the cold gradually became less severe, that the odd few reindeer hunters from the continent started to return. They probably crossed the Öresund Strait by boat from Denmark.

CHAPTER NINE

The Ice Age Ends

A little over 11,600 years ago, the Ice Age came to a clear and definite end. The climate became nearly as warm as it is today. This shift was probably very rapid, with temperature rises of several degrees within just a few decades. Ice cores from Greenland suggest this may even have happened within the space of a few years.

The change was followed in due course by two dramatic events: a gigantic waterfall appeared from nowhere, and, just as suddenly, a strait ran completely dry. It was not until after these two events that people began to move in larger numbers into the territory we now know as Sweden, and to settle there.

Let us exercise a little imagination and picture a few nomadic reindeer hunters travelling in northern Västergötland at that time. Having climbed the mountain we now know as Mount Billingen, they were keeping watch for reindeer herds. It was a warm day in late summer. None of the hunters had ever experienced such a heatwave.

Probably they were puzzled that the reindeer were no longer behaving in the usual way. They must surely have suspected a link with the higher temperatures they were experiencing. However, there was little they could do but stay on the lookout, keep their hopes up, and live on hares and other small animals. Fortunately, there were plenty of fish in lakes and streams, as well as in the salty North Sea. The vast, ice-cold lake to the east, on the other hand, was devoid of life.

By late afternoon, the people on Mount Billingen had still not sighted any reindeer. But they were enjoying themselves anyway. They had enough food for several days, and the streams provided fresh drinking water. Since it was so warm, they no longer needed to muffle themselves up in their usual

thick anoraks and fur-trimmed hoods. A single layer – a thin
tunic of animal hide – was quite sufficient. It was a great
relief to be lightly dressed when making their way up the
mountainside; they avoided sweating as profusely as usual.

They had a magnificent view from the summit. No doubt
their beliefs imbued the whole landscape with spiritual
significance. On this beautiful spot, where they could see
the meeting of mountain, ice, lake and sea, they must have
experienced a particular closeness to deities and spiritual
beings. Bluish ice sheets stretched northwards for as far as the
eye could see. To the east, the dark blue waters of the cold ice
lake extended to the horizon. To the west was the shimmering
sea, with a reddening sun sinking towards the water.

They decided to set up camp for the night and began the
task of lighting a fire. We can picture them sitting in silence,
contemplating in awe the grandeur of the landscape at dusk,
while the fire gradually burned higher.

Then came the first almighty crash. They started, and tried
to speak to each other. But such was the noise that it was
impossible to make out a word anyone said. The thundering
became more intense. It grew worse and worse; this was no
thunderstorm, but the manifestation of other forces that were
far more powerful.

Their eyes widened in terror, they saw immense blocks of
ice calve off the edge of the glacier and plunge into the sea. In
the same instant, a thunderous torrent surged forth, a new
waterfall that had not been there previously. The water
gushed out of the cold eastern lake into the North Sea,
bearing gigantic blocks of ice and boulders so vast that the
combined muscle power of many strong men would not have
been able to lift them.

Frozen with fear, they watched all this while the sun
dipped into the sea and darkness fell. Soon all was black. But
the din continued. They lay down as near as possible to the
fire, and no one moved from the spot all night.

At first light they rose and looked northward to see what
had happened. The new waterfall was even more powerful
and thunderous than the evening before. Great blocks of ice

and boulders were still being swept down to the sea. In the plains to the south-west, a great river had appeared that had not been there the day before.

Getting to their feet, they peered cautiously towards the south. Fortunately, the plains they had come from were still there. As soon as it was light enough, they hurried down the mountainside to tell their kinsfolk about the new waterfall and the new river.

***

Shortly after that, in the area where the Danish city of Helsingør now lies, another small group of people experienced an entirely different kind of miracle.

None of them dared paddle their boats over the narrowest part of the strait. That was far too dangerous, teeming with rapids and waterfalls. To cross to the other side, you had to start your voyage much further north, where the river widened on its way down to the sea and the water was calmer. Yet the most turbulent part of the torrent was still worth visiting around now, in late summer. There were fish to be caught. Above all, with a little luck they might find flints of exactly the right size and the best possible quality.

It was late afternoon on a warm summer's day, the warmest that any of them had ever known. In the morning, the strait looked just as always. As usual, they clambered around gingerly among the stones in their efforts to spear salmon. The least careless movement, and they could be swept away by the cold water, turned upside down and drown.

But that evening, just before sundown, the sound of the rapids ceased abruptly. The rough torrent dwindled into a puny little stream. They sat on the brink, gaping dumbfounded as the water level sank before their very eyes.

By the following day, the water had sunk still further. Within less than one lunar month it had disappeared. The bed of the channel lay exposed. Where lately a torrent had surged through, there was now only fine white sand and pale grey rocks, polished clean. Until lately, even adventurous young people had balked at paddling a boat across the strait. Now anyone, from the tiniest children to the weariest old people, could make their way across on foot. And that was not all. There, on the bed of the channel, lay a treasure trove: endless quantities of the best flint anyone had ever seen. All they had to do was go and gather whatever they needed.

The land we now know as Sweden has been inhabited since before the huge new waterfall appeared and the strait ran dry. Though there are no DNA analyses from the first millennium, we can tell a good deal about the first post-Ice Age pioneers from the DNA of people in neighbouring regions and people from Sweden and Norway who lived 1,500 years later.

**

As the Ice Age drew to a close, people in northern Europe began to change their way of life. Many migrated to the coasts. Reindeer hunters turned to seal hunting.

No doubt the sea had already had a place in our lives at a much earlier stage. The very oldest traces of a culture reflecting modern humans have been found in Blombos cave, on the coast of South Africa. Worked sea shells and leftovers from fish-based meals that are at least 80,000 years old have been found there. As we have already seen, more than 30,000 years ago Europeans were already adorning themselves with shells brought hundreds of kilometres from the Black Sea, the Mediterranean and the Atlantic. Fishing in lakes and rivers, and perhaps on the coast as well, provided a significant complement to the hunting of reindeer, mammoths and wild horses during the Ice Age. One of the very oldest works of art at the French site of Les Eyzies-de-Tayac is a metre-long (3¼-foot) relief of a salmon, chiselled into a rock wall. It is at least 30,000 years old.

But at some point, just as the Ice Age was coming to an end, people in Scandinavia learned how to hunt marine mammals. Initially, they combined the hunting of reindeer inland with seal hunting along the coasts, just as many Inuit people in Greenland and Canada have done for much of history. Some began to make the transition to living on the coast all year round.

Ice Age reindeer hunters probably began to adapt and become seal hunters in Bohuslän (a province on the western seaboard of Sweden, immediately south of the Norwegian border), where the low-lying landscape of the south gives way to the mountains and fjords typical of the Norwegian coast. Unlike the coast of the flatlands, the fjord landscape offers numerous sheltered inlets, which are handy for canoeists who want to land and set up camp.

Some archaeologists refer to the Bohuslän coast of that era as 'the gateway to paradise'. When the Ice Age loosened its grip and the glaciers began to melt, cold, mineral-rich water gushed out into the North Sea, meeting warm saltwater currents. This made the coastal region an El Dorado for shellfish, fish, seabirds and seals – and for people who knew how to exploit this fauna. The Norwegian archaeologist Hein Bjartmann Bjerck believes the transition from reindeer hunting to seal hunting took place in Bohuslän approximately 11,500 years ago, and thinks it was very rapid, taking only a few generations.

At this time, people sharing what appears to be a common culture were migrating across a vast region of northern Europe that stretched from Britain and Belgium in the west, across Doggerland, and as far east as Poland. The arrow points typical of these people are usually assigned to the Ahrensburg culture in Germany, and to the Swiderian culture in Poland. Nearly identical artefacts from the west coast of Sweden are classed as Hensbacka, and as soon as we cross the border into today's Norway they are ascribed to the Fosna culture. The fact that such similar cultures have such different names says more about the narrow purview of twentieth-century archaeologists than about the way people actually lived 11,500 years ago. The hunter-gatherers of the time had no national borders of the kind we know. Rather, they had very extensive networks and travelled over long distances to hunt, meet friends and kinsfolk, hold feasts, exchange gifts and find a partner with whom they could have children.

Doubtless they could walk a few dozen kilometres a day in the open landscape of the tundra. There were as yet no dense forests to stand in their way. The only trees were puny dwarf birches and willows.

The way of life of the Ahrensburgian people initially involved following reindeer on their migrations. When opportunity arose, they also hunted wild horses, elk and bison, as well as fur-bearing animals like wolves and foxes. However, some of them gradually began to see the great advantages of living on the coast for part of the year. It seems likely that they trekked southwards in late summer to engage in reindeer hunting around places like Ahrensburg in north Germany, the area that gave the culture its name. During the rest of the year, their dietary staples were seals, seabirds, fish and shellfish.

To travel along the coast, you needed a boat. Doubtless some kind of relatively simple boats had already existed for a long time. There is evidence that people colonised Australia at least 50,000 years ago, and they must have come by boat. The very first visitors to Sweden, the reindeer hunters of the Hamburg culture who lived 14,000 years ago, are also likely to have had boats; they must have had some means of crossing the Öresund Strait, which was open water at the time. However, it seems likely that the art of boatbuilding in Europe made great advances in Bohuslän, at the very end of the Ice Age.

These craft were probably hide-covered boats similar to those used by the Inuit well into the twentieth century. The tundra that predominated around the end of the Ice Age would not have featured tall trees with thick trunks that could be hollowed out to make boats. But a hide-covered kayak is an equally effective craft that can cover several dozen kilometres a day if paddled by a fit, skilled canoeist. Hein Bjartmann Bjerck believes these boats resembled the Inuit's umiak, an open, relatively large canoe made of sealskin stretched over a framework of laths, which can carry a whole family. There may also have been enclosed kayaks that were smaller and faster, for use in seal hunting.

Once these people had acquired mastery of the maritime environment, they were able to spread northwards along the whole of the Norwegian coast. It seems likely that they simply followed the seals.

The skilled reindeer hunters of the Ahrensburg culture must have found their first seal hunt almost unbelievably easy, trouble-free and productive. These were large animals that lay still along the shoreline to rest, so all you had to do was club them to death. That meant very little effort for copious amounts of fat and nutritious meat. Seals also had a thick layer of fatty tissue that was ideal as fuel, providing warmth and light in the hide-covered tents.

It was almost too good to be true – and sure enough, there was a catch. Seals are smart creatures. They learned to avoid beaches where their kin had been killed by humans, and moved on to new locations. But the people learned how to follow them in increasingly sophisticated boats. While searching for seals, they discovered splendid new dwelling places in sheltered fjords all the way up the Norwegian coast. And everywhere they found the same profusion of fish, shellfish, seabirds and seals.

In the course of a few centuries, Norway's first settlers peopled the Norwegian coastline as far up as the Barents Sea and today's border with Russia. But some people from the new culture of coastal hunters had the privilege of staying on in Bohuslän.

***

During a summer job at *Dagens Nyheter*'s Gothenburg office in the early 1990s, I got to experience the announcement of an archaeological breakthrough. I was one of a group of journalists from western Sweden called to a press conference on the island of Orust, at a place called Huseby Klev. National and local newspapers, radio and television were there. Today, Huseby Klev is surrounded by the wooded farming landscape typical of Bohuslän. But 10,000 years ago, we learned, it was an island far out in the archipelago.

With lively gestures, an enthusiastic archaeologist explained what he and his colleagues had discovered – traces of dolphin hunters who had chewed the world's oldest chewing gum! These finds had been excavated in the course of a survey to prepare for the building of a new road. They had lain deep down and well protected in the clay sediment and were exceptionally well preserved. It had not previously been known that people in Scandinavia 10,000 years ago hunted dolphins. Nor had anyone discovered any 'chewing gum' of comparable antiquity.

Twenty years on I talk to the archaeologist again. His name is Bengt Nordqvist and he recalls that summer day in Huseby Klev as the biggest media event of his life. It was the journalists, he says, who pounced on the phrase 'the world's oldest chewing gum', recognising its potential as a striking headline. Thanks to the chewing gum, Huseby Klev became world news, and Bengt Nordqvist later wrote a short popular science book of the same name.

The archaeologists had found about 10 pieces of resin that had been extracted somehow from birch bark. The tooth marks on these bits of resin show that they were chewed just like modern chewing gum. You can clearly see that it was mostly children and teenagers who did the chewing. In one or two cases an adult finished off the task.

For chewing this prehistoric gum was a task, not just a pleasure. These pieces of resin were put to a practical use. The microscope reveals traces of aspen wood and twisted cords. These fibres, together with the shape of the pieces of resin, show the resin was used for caulking boats. There are good reasons why Scandinavia's first wooden boats were built of aspen wood, aspens being among the first tall, thick trees to appear when the climate began to warm up.

There is a type of traditional boat in the Nordic region made from the hollowed-out trunks of aspens, the Swedish name for which (*esping* or *äsping*) reflects the type of wood used. Bengt Nordqvist believes that the people of Huseby

Klev paddled out to sea in early versions of such craft – long, slender boats carved with great precision out of tree trunks. These boats were low and must have been quite narrow, but they had outriggers to improve their stability. Aspen, a porous wood, was easy to work. However, it split from time to time, and then resin was used to caulk the boats.

Bone remains from the very oldest layers at Huseby Klev show that these people hunted white-beaked dolphins in considerable numbers. These creatures came to the coast in large schools at certain times of the year. The island-dwellers also hunted grey seals, wild boar, red deer and the great auk, which is now extinct. Evidence of cod and ling was found among the fish bones; there were also large quantities of mussel shells.

By the time of the oldest finds at Huseby Klev, the Ice Age was over and conditions were much warmer. Traces of pollen and wood in the excavated layers show how vegetation changed little by little. The first trees were birch, aspen, alder, willow, sallow, rowan and pine. Then came hazel in increasingly abundant quantities. People may have given hazel a little help in establishing itself by pressing nuts down into the soil at favourable spots and felling larger trees that cast too much shadow. After a few thousand years, the environment was dominated by hardwood deciduous trees such as oak, elm and lime.

Once the climate and the seawater had warmed up somewhat, oyster shells were added to the deposits found at Huseby Klev. Similar accumulations of oyster and mussel shells have been found in numerous Stone Age settlements from this time, especially in Denmark. They are known as shell middens or kitchen middens (køkkenmøddinger in Danish). Some are huge – hundreds of metres long and several metres high. We should not be misled by these gigantic waste heaps. While oysters have large shells, they are not exactly filling. A deer provides far more calories than a few handfuls of oysters, but the remains – the bones – take up less space.

The well-preserved finds also show that the people who lived at Huseby Klev ate a good deal of plant-based food. Remains of wild apples, rosehips and sloes have been found.

In particular, there are quantities of shells from hazelnuts roasted in fires. These nutritious nuts seem to have been a staple food; roasting made them taste better and keep longer into the bargain.

A piece of canine excrement has also been preserved in the sediment – a 10,000-year-old lump of dog dirt! The dog itself, however, has not been found. But near Lake Hornborga, in the province of Västergötland, archaeologists have discovered the remains of several dogs from about the same time. The most complete skeleton, which is on display in Falköping Museum, looks as if it belonged to a powerfully built spitz. Several human bones have also been found in the layers excavated at Huseby Klev. Isotopic analysis shows that the bulk of their food came from the sea. It is unusual for people living today to have such an overwhelmingly maritime diet; indeed, it applies only to some of the Inuit of Greenland who retain their traditional way of life.

At the time of writing, the human bones from Huseby Klev are undergoing DNA analysis, but it is unclear whether this undertaking will be successful.

To judge from their bones and the good condition of their tooth enamel, the people who lived at Huseby Klev 10,000 years ago were healthy and well nourished. They seem to have had enough to eat and to have had a balanced diet throughout their lives. 'It must have been the best of all worlds,' says Bengt Nordqvist, summing up.

***

The 'Österöd woman' provides further evidence that people could live well in Bohuslän 10,000 years ago. She was found a few dozen kilometres north of Huseby Klev, next to a croft in the parish of Bro, which is today part of the municipality of Lysekil. At the end of the Ice Age, however, this spot was a small island far out at sea – just a few hours' by canoe from Huseby Klev.

There are some fascinating stories about this skeleton: discovered in 1903 by a quarryman, it attracted the attention of an archaeologist in the 1930s and was described in the light of contemporary knowledge, but then completely forgotten. It was not until 2007 that it was rediscovered by Torbjörn Ahlström, an archaeologist and osteologist, at a museum in Lund. Even more fascinating is what the bones can tell us about the Österöd woman and the kind of life she led.

She was probably about 1.7 metres (5⅗ feet) tall and lived to about 85. Just before she died she was still essentially in good health, free of caries and the osteoporosis that plagues so many elderly women today. The ailments visible from her skeleton include some minor age-related changes in her vertebrae and joints. Her teeth were also rather worn. Such wear and tear is observable in nearly all Stone Age people, who constantly used their teeth as tools when working. There is no evidence that the Österöd woman had suffered from starvation or been seriously ill earlier on in her life.

There are, however, several signs that she engaged in strenuous activities. She appears to have had periostitis, an ailment that can afflict modern-day people who go in for too much jogging or running. She also has a damaged pubic bone, sometimes a sign of having given birth with complications many times, but which can also be caused by playing football and other strenuous sports. Hunting in Bohuslän 10,000 years ago was no doubt at least as dramatic and violent as a modern game of football.

Isotopic analyses of the woman's teeth show she must have eaten a great deal of venison and elk meat from mainland forests, in her childhood at any rate. In this she differs from the people who lived at Huseby Klev, who lived almost exclusively on fish and marine animals throughout their lives.

The Österöd woman may have belonged to a group that migrated from season to season between the hinterland and the outer archipelago. She may also conceivably have spent her childhood in the forests of the mainland, but moved to the coast as an adult and adopted a more maritime way of life.

Torbjörn Ahlström shares the conclusion that life in Bohuslän just after the end of the Ice Age must have been a very special time – 'a real bonanza', as he says. I am ready to agree with him. Obviously we should avoid romanticising life during the Stone Age. But at that time Bohuslän was a virgin territory with an abundance of food, and there were so far very few people to compete for resources.

Even today, the west coast of Sweden can be a harsh place when it is at its rainiest and windiest in autumn and winter. The hide tents of the dolphin hunters probably afforded rather inadequate shelter in such weather. On calmer days, however, Bohuslän can be like paradise. Ten thousand years ago, just like today, there must have been still mornings at the height of summer when the sea was as smooth as a mirror; and autumn days when the hazel bushes had turned yellow and the leaves of the aspens red, when the warmth lingered on, with the tang of seaweed and the cries of seabirds in the air. It must have been pleasant then to lie on a slab of smooth pink granite listening to the lapping of the waves, while eating shellfish and nibbling freshly roasted nuts.

Huseby Klev and Österöd compete for the title of Sweden's oldest human remains. There is no consensus among archaeologists, partly because a marine diet makes radiocarbon dating more unreliable than usual. By way of a compromise, let's just say that both sites are about 10,000 years old, with a few centuries' margin of error.

Researchers have attempted to analyse DNA from the Österöd woman, but unfortunately they have not managed to extract enough of it. Analyses of a number of skeletons from southern Norway that are a few hundred years younger have proven more successful.

***

It has been suggested more than once that a few daring seafarers may have taken a shortcut straight across the sea from Doggerland to the coast of Norway. However, Bohuslän is generally viewed as the most likely point of departure of Norway's first settlers.

A number of archaeological finds point to another migratory route, thought to have started in the east, in Russia — in fact, even as far away as Siberia. Most of these finds are a kind of stone arrow point made using a particular technique that involved the toolmaker pressing down on the stone, rather than knapping it. Arrows of this type were made early on in Siberia and have also been found in Russia and Finland. Finds of such arrowheads indicate that people migrated towards the Arctic and possibly also along the southern shores of the Baltic Sea. Bone fish-hooks fashioned in a specific way tell the same story.

The question is: what can DNA analyses tell us? The oldest and best-researched representative of Ice Age Siberians lived 24,000 years ago at a place called Mal'ta near Lake Baikal. He was about four years old when he died, and to judge from DNA analyses he probably had brown hair, brown eyes and a pale, freckled face (though the DNA analyses are somewhat inconclusive on these points).

When this information about the Mal'ta boy's DNA appeared in *Nature* in 2013, it caused a big stir. This was because he turned out to belong to a group of people who were midway, genetically speaking, between western Europe's early population of hunters and the indigenous people of the Americas. In other words, the 'ancient Siberians' formed a kind of missing link between Ice Age Europe and the Americas of the same period. There is no extant group of people with an identical DNA set. However, many of us, both in the Americas and in Europe, have inherited some of the ancient Siberians' genetic material. These sequences of our DNA can provide insights into prehistoric migrations.

The indigenous people of the Americas are descended from groups that lived in Beringia, a region that, like Doggerland, no longer exists. It now lies on the seabed around the body of water known as the Bering Strait. Beringia covered a very large area, far larger than Doggerland, and new DNA research indicates that the indigenous people of the Americas left it in waves, starting over 14,000 years ago.

Now researchers are looking for a similar DNA signal from the east in the oldest individuals in Scandinavia to have been subjected to DNA analysis. They are about 9,500 years old and were found at a place called Hummerviken, in the municipality of Søgne on the western side of the Oslo Fjord, Norway.

In the 1990s, the owner of a cottage on a plot of land running down to the shore was planning to carry out dredging work in preparation for installing a new landing stage. In the clay on the shoreline, a human bone was found – the oldest ever discovered in Norway. It was suspected that the individual was a woman, and the mass media began to call her 'Sol' (Sun). Later, more bones from a few other individuals were found. They appear to have died at the same time, a reasonable assumption being that they all drowned together when a boat foundered.

The DNA of two of these individuals has been analysed by Swedish researchers; the others are being studied by a rival team whose leadership is based in Copenhagen. The results available so far show that the people from Hummerviken had mitochondria from haplogroups U5a1, U5a1d and U4d. The same haplogroups have also been identified in individuals found in Sweden who lived just a few hundred years after the Hummerviken people.

***

Boats crossed the Baltic as well. On the island of Stora Karlsö just off Gotland, Sweden, archaeologists have found traces of humans in a cave known as Stora Förvar. These finds reveal that people were already coming to the cave some 9,400 years ago, after which they continued to visit it for thousands of years.

The main reason for their presence was probably to hunt grey seals. The layers in the cave contain large numbers of seal bones, almost exclusively from cubs. Presumably they would have begun hunting seals towards the end of winter, sometime in March, and continued for a few months. Later in the season they would have focused on other prey, such as hares and seabirds, as well as on fishing and picking hazelnuts.

Archaeologists have found bones and skulls from about 10 individuals in the Stora Förvar cave. One of the skeletons is that of a baby who died at about four months of age. There is a theory that the child was poisoned by overdoses of vitamin A, a possible consequence of its mother consuming too much seal meat – especially liver – while breastfeeding. Such meat is very rich in vitamin A, which is vital in small quantities but toxic in excessive amounts, especially to young children.

Hunting peoples heavily reliant on marine mammals in modern times – such as the people living along the west coast of North America – have learned that pregnant women and nursing mothers have to eat other kinds of food. But archaeologists suspect that the people on the island of Stora Karlsö were beginners. They had only recently made the transition to maritime hunting, having previously hunted mainly deer and elk in the forests on the mainland.

The dead in the Stora Förvar cave also include a few children aged about 10, at least one of whom was probably a girl. Two or three individuals were adolescents, while the rest were in their twenties and thirties.

There are various theories about how these human remains found their way to the cave. Perhaps they were people who just happened to have died while the group was on the island of Stora Karlsö. Or they may have died during the year and been brought to Stora Förvar for burial because the cave was believed to have a special significance.

Among the remains of seal bones, hares, seabirds and fish bones, there are also traces of a type of prey that could hardly have been used as food. The people on Stora Karlsö were eagle hunters. We can only speculate about the exact use to which they put their prey. However, it seems natural to assume that these great, powerful birds would have had a symbolic significance, and that their plumage was used to decorate people's clothing.

I have in mind a number of pictures of people from North America's indigenous population taken by the photographer Edward Curtis in the first decades of the twentieth century. Several of them are wearing traditional headdresses made of large feathers. The seal hunters of Stora Karlsö may have had a similar appearance. They may have worn sealskin and deerskin garments adorned with animal teeth and shells, with long eagle feathers in their hair.

A Swedish research team has analysed the DNA of several of the individuals from Stora Förvar. The oldest lived about 9,200 years ago and, at the time of writing, was the oldest individual in Sweden to have been subjected to DNA analysis. This person belonged to haplogroup U4, the same found in Hummerviken.

The people of Hummerviken and the inhabitants of Stora Karlsö probably belonged to the same related group of hunters. To judge by their mitochondria, they may actually have been descended from people further east. Their foremothers may have spent the coldest period of the Ice Age in the southern part of Ukraine or Russia, or even in western Siberia – at least according to analyses of people alive today carried out by Russian researchers and others.

It is to be hoped that more detailed analyses of nuclear DNA will soon give us more information. Maybe they will provide some clues about where these early Scandinavians came from and what they may have looked like.

CHAPTER TEN

Dark Skin, Blue Eyes

The Barum woman lived just over a millennium after the woman from Österöd and 500 years after the first seal hunters on the island of Stora Karlsö. But she is far more of a celebrity.

She is exhibited in the Swedish History Museum in Stockholm, where she used to be known as the Bäckaskog woman. The name has been changed because archaeologists and locals in Skåne, the southern province where she was found, have always called her the 'Barum woman' after the village where she was discovered.

One reason why the Barum woman is so much better known is that her skeleton is so well preserved. Visitors to the History Museum can view her in a glass case, seated in approximately the same position as at the times of her burial and discovery (though archaeologists disagree on certain details, such as the angle at which she was originally leaning).

When the skeleton was first discovered in 1939 at Lake Oppmanna, just outside Kristianstad in Skåne, the archaeologists assumed they were looking at a man – admittedly a very short one at barely 1.55 metres (5⅛ feet), but definitely a man. After all, he had been buried with a bone arrow accompanied by flint arrowheads and a chisel made of elk bone – gifts that were clearly suited to a male. He hit the headlines as 'the Barum fisherman'.

However, in the early 1970s the skeleton was re-examined and an expert at the History Museum identified some typically female characteristics. One of these was the same kind of damage to the pubic bone as that identified in the Österöd woman from Bohuslän. They were interpreted, in accordance with the knowledge of the time, as showing that she was a woman who had borne about 10 children. Moreover, isotopic analyses showed that she had by no means lived on fish alone, but rather on deer, elk and other forest-dwelling mammals. Viewed in conjunction, the analyses of her bones and teeth suggest that she was aged between 35 and 40 at the time of her death.

In the information provided by the History Museum, the 'Barum fisherman' was transformed into a hunter and mother of 10 – a real Swedish matriarch, perfectly in tune with the seventies' zeitgeist. However, according to more recent insights from sports medicine, the damaged pubic bone may not necessarily signify that the Barum skeleton belonged to a mother of 10. She may have had fewer children but have engaged in strenuous activities as a hunter.

The History Museum has also had a model-maker reconstruct the Barum woman's appearance on the basis of her skeleton. She and I are about the same height. It is quite an experience to stand face-to-face with a person who lived nearly 9,000 years ago. She looks almost uncannily real; the sculptor has done a skilled job based on the archaeologists' brief. He has made her light-skinned, with greenish eyes and hair of an indefinite, but fairly light colour, with a sprinkling of grey. Her hair is lank, tied back simply at the nape. She has no embellishments or adornments.

I can't help wondering whether she was really so plainly attired. What I have learned so far about Stone Age people is that they left plentiful traces of jewellery and pigments. I have in mind a woman from Bad Dürrenberg in Germany, who lived at about the same time as the Barum woman and whose skeleton is displayed in the Museum of Prehistory in the German city of Halle. There, by contrast, the scientists told their illustrator to lay the decorations on thick, including animals' teeth, wild boar tusks, feathers and stripes of red pigment.

Another question is whether the Barum woman really was as light-skinned and light-haired as the model in Stockholm's History Museum. She may very well have had both darker skin and darker hair. However, she probably had blue eyes, no matter what colour her skin was.

***

The vast majority of individuals from the hunting Stone Age in Europe whose DNA has been analysed belong to mitochondrial group U. Many of them, like me, belong to U5. Some – again like me – belong to subgroup U5b1. This is true of the two individuals from Bonn–Oberkassel who were buried together with a dog 14,500 years ago. The same applies to their near relative from Loschbour, Luxembourg, who lived about 8,000 years ago. He was in his forties when he died, 1.6 metres (5¼ feet) tall, weighed about 60 kilos (132 pounds) and was interred with red ochre scattered over his head.

There are also DNA analyses of two people found in the Spanish cave of La Braña, near the Atlantic coast of northern Spain. These two individuals also lived nearly 8,000 years ago and belong to mitochondrial group U5b, but they are part of a different subgroup known as U5b2. The one most thoroughly analysed was a well-built man in his twenties. The drawings published by the Spanish research team along with their scientific article depict him with black hair and dark skin, more or less like an African. Yet his eyes are light blue.

Blue eyes and dark skin are an extremely rare combination today, but many Ice Age people seem to have had such an appearance. Blue eyes, dark hair and dark skin seem to have been commonplace in much of Europe. There is a degree of uncertainty as regards genetic analyses of human pigmentation. The level of probability is lower than the norm in scientific studies published by scientific researchers. Moreover, the colour of people's skin and hair is influenced by dozens of genes. However, a number of gene variants have a particularly decisive impact, and several studies published by rival research teams, all pointing in the same direction, are now available.

Quite simply, it looks as if many people among Europe's original population of hunters had black hair and quite dark skin. It is self-evident that the people who first arrived from the Middle East looked like that; after all, they had come from Africa, where everyone was dark. But as late as 8,000 – or even 5,000 – years ago, many European hunters still bore the original gene variants from Africa, meaning that they were probably quite dark-skinned.

A few millennia later, nearly all Europeans were born with light skin. The original gene variants from Africa had undergone a transformation. There must have been factors at work that enabled light-skinned people to thrive better in Europe's northerly latitudes, to survive in greater numbers and to have more children who survived in their turn.

The most obvious mechanism has to do with the sun and its ultraviolet rays. Too much sun is dangerous, resulting in burns, skin cancer and, probably, reduced immune defence. That was why the forerunners of humans in Africa developed dark skin as soon as they had lost their protective covering of hair, millions of years ago. But when some people began to migrate northwards to latitudes where the sun's rays are weaker, they were at less risk of being burned. Indeed, their dark pigmentation could even pose problems by preventing the skin from producing vitamin D.

Vitamin D deficiency is a serious condition. It is one of the causes of rickets, which results in poor-quality teeth and crooked bones. In particular, it leads to malformations of the pelvis, which must have been devastating for young women in prehistoric Europe when they had to give birth for the first time. There was a major risk that both the mother and the baby would die. If that happened, the woman's genes were not passed on to the next generation. This must have resulted in a rigorous process of elimination, with light-skinned people whose skin could readily produce vitamin D passing on their traits in greater numbers.

This was not necessarily a major problem for Europe's early population of hunters, many of whom ate large quantities of fatty fish, with coast-dwellers consuming meat and liver from seals and dolphins as well. This enabled them to meet their vitamin D needs. However, for those who lived in the far north – in today's Sweden, for example – and lived chiefly on plants and inland mammals, lighter skin may have been an advantage.

With the transition to agriculture, the pressures that favoured light skin increased throughout Europe. The food eaten by farming people was generally poorer in vitamin D than the hunters' diet. So two complementary factors seem to have played a role in gradually lightening the once-dark skin inherited from African forebears: the sun was weaker in northerly latitudes, and people's diet changed as farming took hold.

***

Individuals who lived nearly 8,000 years ago were found in Motala, in the Swedish province of Östergötland, and subjected to DNA analysis. They were discovered in the course of a few excavation seasons in 2009 and 2010.

The excavation site, known as 'Kanaljorden', lies next to the River Motala. Eight thousand years ago the site was strategically positioned on the shores of Lake Vättern, next to a torrential watercourse that led down to the sea. It was easy to reach the spot by boat from a number of directions, making it an ideal venue for feasts.

In total, archaeologists have found the remains of about 15 people. However, oddly enough, there are only a few bones from each individual – specifically, their skulls. These skulls had been placed on the bed of a small lake just a few metres deep. They were found lying next to some large boulders that had been piled together to form a kind of platform.

The archaeologists have only been able to find a whole skeleton for one individual – a newborn baby. Vestiges of wood reveal that the adults' skulls were mounted on stakes. A wooden art object has also been found, a large fish carved out of a plank. This, too, appears to have been mounted on a pole.

Why did these unfortunates end up on the lake bed? One interpretation is that they were the victims of a conflict; their enemies, who emerged victorious, stuck their heads onto long stakes as trophies. After all, they died at a time when competition for resources seems to have intensified.

However, the archaeologist Fredrik Hallgren, the former project leader at the site, prefers another explanation. His theory is that small groups of people from different areas met once a year on the River Motala at a collective encampment. This would have happened at a time when there was ready access to sufficient food to keep all the participants fed for a few weeks. It was probably in spring, when the sea trout were making their way upstream. Family and acquaintances from a widespread network seized the opportunity to meet. They fished and feasted together, exchanged information and gifts, socialised, and looked for a partner. And they also held a solemn ceremony to bury their dead.

Adults who had died in the course of the year were too heavy to be carried all the way to the camp; that was why only their skulls were brought there. The small body of the baby, however, was light enough for its parents to carry. Most of the skeletons seem to have been clean. Either the relatives of the dead people had scraped off the flesh with knives, or the bodies had been allowed to decompose in the ground until they were exhumed and the bones were taken to the great ceremony. In one case, however, a head seems to have been placed in the lake just a short time after the individual's death, as vestiges of the brain are still preserved inside the skull.

Fredrik Hallgren imagines that the skulls and the large wooden fish would have been set on poles over the surface of the water and that they would have played an important role in the ceremony.

In my mind's eye, I picture these people holding their ritual early in the morning, when the sun was veiled in a thick mist. There would have been fires burning on the shores. They would have sung in honour of the dead, played instruments and danced.

Archaeologists have examined isotopes in the bones that show the type of bedrock in the regions where these people lived. Apparently none of them come from Motala, where the skulls were found. Rather, they spent most of the year in three separate places up to 50 kilometres (30 miles) away. Some of them came from the forests of Södermanland to the north, others from the forests to the south and the Baltic coast to the east. Some of the women seem to have moved at around 20 years of age, which probably indicates that they met a man at that time and moved to join his group.

An international team of researchers has helped to conduct DNA analyses of seven individuals from Motala. These analyses show that they belonged to haplogroups U5a1, U2e1, U5a2 and U5a2d. The researchers have also examined their nuclear DNA, which provides far more information than mitochondrial DNA alone. The results show that the people found in Motala were quite closely related to other Stone Age hunters from all over Europe and even from distant Siberia. They are most similar to other Stone Age hunters from the same region – those found on the islands of Stora Karlsö and Gotland, the oldest of whom is 9,400 years old, while the youngest dates back about 5,000 years.

At the time of writing, the DNA of over 20 Stone Age hunters from Sweden has been analysed. The genetic results suggest that the vast majority had blue eyes. Some of them still had the original genes from African times that make for a dark skin. However, the majority – about two out of three – had acquired the gene variants that produce a fairer complexion.

So most of the people living on Gotland and in the Motala region during the hunters' Stone Age had lighter skin than their close relatives on the Atlantic coast. There was variation, just as there is today; some were a little darker, others a little lighter-skinned. On average, the hunters of that time appear to have had somewhat darker skin than modern Swedes; besides, they were outside all day, so they were more tanned than many Swedes today.

Many of them also had a specific gene that is common today in China and other East Asian countries, as well as among the indigenous peoples of the Americas. This gene is believed to be responsible for strong hair with thick strands, comparatively small breasts in women, and front teeth with a particular shovel shape, common among Chinese people.

As for me, I have blue eyes, very blonde hair and very fair skin – like the stereotypical Swede. But the truth is that far from everyone in today's Sweden looks like me. There are plenty of people with dark hair and brown eyes who can trace their origins within the Nordic region back hundreds of years. And now DNA research suggests that both dark hair and dark skin have been here ever since the ice sheet melted and the first people arrived here.

CHAPTER ELEVEN

Climate and Forests

During the lifetime of the Barum woman, temperatures in Scandinavia were more or less as they are today. The forests were dominated by pine and hazel. Nutritious hazelnuts were a key part of people's diet. Hardwood deciduous trees such as oak, elm, ash and lime were also to be found in the forests. About 8,800 years ago the climate started to become damper and slightly warmer, which was favourable for hardwood trees.

When the people of Motala gathered some 8,000 years ago for their strange funerary rites, the southern half of Sweden had a climate not unlike that of England today. The woods became ever denser, with ivy and prickly bushes occupying the ground between the trees. There were plenty of dead trees to provide a habitat for birds and insects. This type of natural environment dominated western Europe for thousands of years, yet today there are virtually no traces of it left. Practically all the ancient deciduous forests have been cut down to make way for farmland, towns and roads.

To experience a forest something like those where our relatives lived during this part of the Stone Age, you have to visit a nature reserve. In such surroundings, you can find small fragments of the original European deciduous forest. One example is Vitsippsdalen ('the Vale of Wood Anemones') in Gothenburg's botanical gardens, where I walk whenever the opportunity arises. When I am there, I think about how humid, leafy woods just like this, full of songbirds in spring, were long the norm in the southern half of Sweden and across much of Europe.

The most extensive and cohesive vestiges of Europe's virgin forests are to be found in Białowieska, on the border between Poland and Belarus. When travelling through the region a few years ago, I was struck by the impenetrable density of the forest and by how difficult it is for the sun to filter through the leafy treetops. When dense forests of this kind took over in southern Scandinavia, the habitat was no longer suitable for reindeer, horses or the great aurochs. Deer and wild boar accounted for most of the available quarry. On the one hand, such game animals were plentiful, as the forest provided such a wealth of food. On the other, it was a challenge for people to locate their quarry in such dense vegetation, and actually moving around within the forest was difficult.

Archaeological finds from Denmark and elsewhere show that bows and arrows became common at precisely the time that forests were becoming denser. Spear throwers had been an effective weapon in the earlier open landscape characteristic of the Ice Age. But now that people needed to pick their way stealthily through dense forests and target their prey from different angles, bows were a much better weapon. And of course there were dogs too. In my view, the archaeologist Martin Street, based in Germany, is on the right track with his idea that hunting with dogs came to play a more important role once the open landscape was replaced by dense forests. Dogs may also have helped people hunting waterfowl from canoes by swimming to retrieve their prey.

Canoes were an indispensable means of transport when the forests grew denser. On land, all the trees and bushes made it more difficult to travel long distances, as people had been able to do when the landscape was more open. Moreover, people gradually felt less need for contact with groups living a long way away. The good life in virginal northern Europe in the warmer period after the Ice Age gave a big boost to population size, thereby improving the chances of finding companions and marriage partners close at hand. There was no longer any need to trek over hundreds of kilometres to have some company – two or three dozen were sufficient.

As a result, groups of people became more isolated. This can be seen from the tools they left behind; increasingly, these show specific local features.

***

When the climate was at its warmest, Sweden was a few degrees warmer than it is today; temperatures were more or less what they are expected to be in a few decades' time, as a result of the global warming that is currently under way.

But 8,200 years ago there was a backlash. The cold made a comeback, probably owing to the weakening of currents in the Atlantic, including the well-known Gulf Stream. This brief cold spell lasted for a few centuries. It was at its worst for a period of about four years. People who had managed to adjust to a mild climate were catapulted back into a climate as bitter as that of the Ice Age. It was as if modern-day people from southern Sweden suddenly had to cope with a climate like that of Greenland.

As if it were not enough that temperatures varied unpredictably, the same applied to the sea level. When the great glaciers of the Ice Age melted, the water ran out into the sea and other bodies of water. Moreover, the land rose in much of Scandinavia. The bedrock, once weighed down by kilometre-thick layers of heavy ice, rose up again once the masses of ice had disappeared.

The land is continuing to rise in much of Scandinavia, but it was a much faster process initially. The hunters of the Stone Age sometimes witnessed entirely new land masses emerging from the sea within a matter of years. Conversely, their settlements and habitual hunting grounds were often engulfed by water.

That is exactly what happened to Doggerland, where the flooding was particularly catastrophic. People had lived for millennia in the regions between Britain and Denmark. There may have been relatively small groups of mammoth and reindeer hunters there even during the Ice Age. Unlike Scandinavia, which was largely covered by thick glaciers, parts of Doggerland seem to have remained ice-free even during the coldest periods.

After the end of the Ice Age, Doggerland's climate grew warmer, just as in the rest of northern Europe. The vegetation changed. Hazel bushes spread over an increasingly wide area where once there had been only tundra, and gradually dense forests of hardwood broadleaf trees also emerged. The old fauna that people had once hunted disappeared. The people who remained in the region were obliged to adapt to a new way of life based more on hunting deer and wild boar.

At the time when the climate was at its very coldest, parts of the northern coast of Doggerland probably stretched up to the same latitude as the Shetland Islands and the Norwegian city of Bergen. However, the more the sea level rose, the more Doggerland shrank. Researchers calculate that the sea rose by 1.25 metres (4⅛ feet) per century. That corresponds to the changes today's young people may experience if our emissions and global warming continue at their current rate.

Doggerland dwindled, becoming increasingly dominated by river deltas, lakes and marshes. Covering reasonably long distances on land became more and more difficult. Canoes became an essential means of transport. Increasingly, the hunting of deer and wild boar was supplemented by fishing and hunting seabirds. The whole region underwent major changes. For all that, however, it was a remarkably productive landscape; some scientists describe it as one of the most favourable environments in the whole of Europe.

The cold spell that occurred 8,200 years ago was, of course, a major blow that doubtless caused many deaths, both in Doggerland and in the rest of north-western Europe. Three or four generations later – 8,100 years ago – disaster struck once more. Large tracts of rich countryside disappeared under water.

The reason for this was the Storegga Slide, a very extensive landslide on the seabed in the Norwegian Sea between Iceland and Norway, at the same latitude as the Norwegian city of Trondheim. This landslide gave rise to a tsunami that was the most powerful ever documented in northern Europe.

Along the Norwegian seaboard, the land closest to the slide, the waves were between 10 and 12 metres (33 and 39 feet) high. No one who was near the shore at the time can have survived. However, those that were up in the mountains – on a hunting expedition, perhaps – would have managed to stay alive. One can only try to picture these hunters' horror when they returned and found their base camp destroyed by the waves, with all their family members drowned.

No one on Doggerland's low-lying shoreline can have survived either. Although the tsunami waves were only three or four metres (10 to 13 feet) high here, the land was flatter than the Norwegian coast, and there were no high mountains where anyone could take refuge.

It is, of course, theoretically possible that a few of Doggerland's inhabitants were in their canoes far out at sea, perhaps to fish for mackerel or hunt seabirds. Out on the deep, the tsunami would have been scarcely perceptible, just a few centimetres high. The boat would have rocked a little, but the seafarer would not have felt anything else – until he or she paddled back to the shore, where the wave had destroyed everything. The whole settlement would have been flooded and all friends and family swept away.

The sea level continued to rise over the centuries that followed. Doggerland is still under water. The land that was home to people for millennia now lies at the bottom of the sea.

***

I began to ponder when my own foremothers had arrived in Scandinavia and which route they had taken. Did they paddle across the sea from Doggerland? Did they walk across a dry Öresund Strait, or did they take the long way round via Finland? To answer these questions, I needed more comprehensive analyses of my mitochondrial DNA than the Icelandic company deCODE was able to offer.

In the summer of 2013 I bought a test kit from Family Tree DNA (FTDNA), a US company that specialises in catering to family history researchers. I scraped the inside of my cheek with a small plastic spatula, put the sample in a test tube and sent it in an envelope to the laboratory in the United States. A few weeks later I received the full details of my mitochondrial DNA – the most exhaustive mitochondrial test in existence.

I found out that I belong to subgroup U5b1b1.

CHAPTER TWELVE

Am I a Sami?

My first reaction when I get to see my result is: 'Really? Do I have Sami ancestry?' It's a big surprise. Nothing I've heard so far about my family history has given any indication of such a link. My oldest known female relative in the direct maternal line lived in the western province of Värmland. No one has ever mentioned that she may have been a Sami.

But the mitochondrial group U5b1b1 is particularly common among Sami people. Nearly half of all the Sami whose DNA has been tested belong to this group. Their particular genetic variant of mitochondria is even known as 'the Sami motif', and it is rare among other Europeans.

It would be cool to be able to prove some Sami roots, I think to myself. Not that it would have any practical or legal implications, of course – any family connections would be much too far back in time.

The Sami are classed as Europe's only indigenous people. That means they have special rights laid down in international conventions. All adult Sami are entitled to vote for the Sami Parliament (*Sámediggi*, or *Sametinget* in Swedish). Those who belong to Sami villages also have certain privileges as regards hunting and fishing.

To enjoy rights like these, I can't just come along with a DNA sample that proves there were family links several centuries ago. I would have to demonstrate some Sami traditions in my immediate family. At least one of my parents or grandparents would have had to speak a Sami language at home. Being adopted – without any genetic links at all – is also accepted for purposes of recognition.

Tens of thousands of Scandinavians with Sami roots have the Sami motif. Their common foremother probably lived in northern Spain or southern France 18,000 years ago, when the Ice Age was at its coldest. When it grew warmer, her descendants followed the reindeer northwards, but their route seems to have included a diversion along the way.

In 2004, the Estonian researcher Kristiina Tambets published her doctoral thesis and an article in a scientific journal, entitled 'The western and eastern roots of the Saami'. One of the people with whom Tambets collaborated was the late geneticist Lars Beckman, from Umeå, who worked on genetics in northern Sweden for several decades, beginning in the late 1950s. While he focused mainly on the genetics of disease, he was also interested in the question of how Norrland (the northern part of Sweden) was first populated. Today's DNA technology was not available to Beckman. He worked mainly on individual markers, such as blood groups.

However, he did manage to publish a number of DNA studies in cooperation with other researchers before his death in 2005. With the help of DNA from samples collected by Beckman, Tambets and other researchers were able to map the route taken by some of the female ancestors of today's Sami people. They probably left northern Spain and southern France when the climate grew warmer after the Ice Age and migrated northwards. But then they changed direction. They took a turning and continued eastwards – straight through central Europe towards Poland, Belarus and Russia. From there, they gradually migrated further north, via Finland.

This is why there is considerable variation in haplogroup U5b1b1 in Spain and France, while it is also dispersed through countries such as Slovakia, Poland, Russia and Belarus. It is also to be found even in countries such as Morocco and Senegal, where the descendants of people who migrated south rather than north live today. In much of Europe, haplogroup U5b1b1 is very rare – under 1 per cent. Other haplogroups have become far more common thanks to subsequent waves of immigration. It is only in the very far north that the rare DNA variants are still to be found on a larger scale. An even rarer haplogroup known as V, which also occurs among Sami people, seems to have a similar history, originating in south-west Europe and moving into the eastern and northern regions of the continent.

We should bear in mind that the Sami motif U5b1b1 and haplogroup V only tell us a limited part of the story about Sami origins. Like most other ethnic groups, the Sami are a mix of people from different places.

The Uppsala-based geneticist Ulf Gyllensten published a study in 2006 in which he showed that haplogroup Z, which also occurs among Sami people, is most likely to have originated near the Urals and the Volga, in Russia. While U5b1b1 seems to have existed in Finland and Scandinavia for at least 6,000 years, Z apparently arrived at a later stage. That suggests that there was a new wave of immigration from the east only a few thousand years ago.

To find out more about the earliest history of the Sami, I take the night train up to Jokkmokk, in the province of Lapland, and meet Kjell-Åke Aronsson, the director of the Sami Museum, Ájtte. A trained archaeologist, he wrote his doctoral thesis on how reindeer were domesticated for human benefit. He and others conclude that domestication took place in comparatively recent times, perhaps some 1,500 years ago during the Iron Age. There were wild reindeer in northern Scandinavia before that time, but they were not a dominant factor in people's lives. Fishing and hunting elk were far more important means of subsistence.

The natural environment in northern Sweden was not ideal for reindeer until large spruce forests began to expand there. This happened only a few thousand years ago as the climate became somewhat colder. Reindeer like to winter in spruce forests because they can graze throughout the season on the lichens that grow on spruce pines. Deciduous forests, by contrast, offer little for reindeer to live on, especially when the temperature oscillates around zero degrees Celsius (32 degrees Fahrenheit). A hard ice crust forms at such temperatures, denying the reindeer access to lichen on the ground. Intense cold is less of a problem, as snow is soft and fluffy under such conditions, making it easier to scrape off.

Aronsson's work on his doctoral thesis was based mainly on analyses of pollen. These showed how vegetation had changed as domesticated reindeer became common. This was because reindeer droppings acted as a fertiliser, while their grazing habits also affected the vegetation. More recently, Aronsson has begun to examine reindeer DNA in collaboration with Norwegian researchers. However, they have not yet succeeded in identifying any wild reindeer with DNA matching that of domestic animals, which would reveal the latter's origins. They are pinning their hopes on bones and teeth from the ancient wild reindeer that are emerging from glaciers as the climate grows warmer.

The Sami languages, too, are a comparatively recent phenomenon. They are Uralic languages related to Finnish, Hungarian and a number of minor languages spoken in northern Russia, especially in parts of the Urals. Linguists have found indications that the Sami languages developed near areas where Baltic languages were spoken – that is, further east than the areas where Sami languages are spoken today.

Aronsson has his own ideas about how the Sami languages came to northern Scandinavia. His theory is slightly controversial – 'like swearing in church', he says. After a little persuasion, he agrees to tell me it anyway. He sees a link between the Sami languages and the early use of iron. These two phenomena appear to have reached northern Scandinavia at around the same time, about 2,000 years ago. If a small group of immigrants were able to transfer a whole new language to an existing population, they must have had something very special to offer, Aronsson thinks. In his view, domesticated reindeer are not a sufficient explanation. Conversely, the art of using fire to transform stone and sand into gleaming metal would be a good reason.

He also believes that aspects of the traditional Sami religion accompanied the language and the art of metalworking. The Sami shamans, known as *nåjder*, often decorated their drums with metallic objects. Pewter and silver are important components of traditional Sami jewellery and costumes.

Aronsson's theory is based in part on recent archaeological excavations, including some carried out before the building of the railway line between Boden and Haparanda. In the course of the dig, archaeologists found objects several thousand years old that bear witness to a new type of settlement where iron was produced.

My own view is that the Sami languages, metalworking and a number of people may have arrived from the east around 1,500 to 2,000 years ago. This pattern would agree with Ulf Gyllensten's findings that mitochondria belonging to haplogroup Z also appeared at around that time. Moreover, a large proportion of Sami men's Y chromosomes reveal kinship with groups living further east. Nearly half of all the Sami men whose DNA has been analysed belong to group N3, which is common in Siberia and in groups speaking Finno-Ugric languages. In western Europe, by contrast, this haplogroup is a rarity.

When I look into the issue more deeply, it turns out that my mitochondria lack the specific mutations known as the Sami motif. These mutations form a special subgroup of U5b1b1 called U5b1b1a. I, on the other hand, belong to another, far more unusual little group.

Initially, the administrators at FTDNA manage to identify just five people in the whole world who belong to the same small group as myself. Two of them are Americans with Norwegian roots, which makes sense, given that my oldest known female ancestor came from western Värmland, just a few dozen kilometres from the Norwegian border. The two Americans appear to be closely related, even though they have not found any written sources with supporting evidence. But their lineage and mine separated several thousand years ago.

Three of my matches come from Spain and Portugal, no less. That seems a little puzzling, but can be explained by the fact that haplogroup U5b1b1 originated in that region during the Ice Age. Many of us migrated to the north, but there were others who stayed put.

A few months later I get another bite. There turns out to be a woman in Stockholm whose mitochondrial DNA is closer to mine than that of the Spaniard and the two Americans. There are a few steps between us, which may also correspond to a few millennia. Her husband, an active family history researcher, sent the sample to FTDNA. On hearing that we are related, he takes a look in some church records and manages to trace his wife's oldest relative in the maternal line, one Margareta Nilsdotter, who died in 1687 in Burträsk. Burträsk is a small place in Västerbotten, one of whose claims to fame is that the world's oldest skis were found nearby – the 'Kalvträsk skis', which go back 5,200 years.

Burträsk! So there may be some Sami roots here after all, I think again. However, nothing in the church records suggests that Margareta Nilsdotter might have been a Sami.

My particular branch and the lineage of everyone with the same variant went their separate ways several millennia ago – perhaps 10,000 years ago, although there is a considerable margin of error. The female forebears of the Sami and my own foremothers seem to have lived near the Nordic region for many thousands of years, possibly even since the ice sheet melted. This is fascinating for a family history researcher, but it is important to remind yourself that these issues are of historical interest only. Who went in one direction or another several thousand years ago has no bearing on today's Sami identity or on the issue of the Sami's status as an indigenous people. 'Being Sami is not a genetic thing,' Kjell-Åke Aronsson emphasises.

The fact that the Sami are classed as an indigenous people by a UN convention springs from events over the last few centuries. It was during this period that the Swedish state occupied the regions that Sami groups regarded as theirs. They resorted to brutal methods at times, causing great and enduring bitterness.

In the seventeenth century, the main problem – from the perspective of the Swedish state and the Protestant church – was that the Sami were not Christians, explains Aronsson. They were regarded as heathens. The Swedish state sent them missionaries, priests and other representatives, who confiscated the shamans' most vital piece of equipment, their magic drums, and destroyed many votive sites. One shaman was executed for witchcraft.

In the eighteenth century, relations between the Swedish state and the Sami improved somewhat. Linnaeus writes enthusiastically about his encounter with the Sami in his book *Lachesis Lapponica: A Tour in Lapland*. In particular, he notes how healthy and vigorous elderly Sami people were in comparison with the Swedish-speaking farming population.

But taxes were burdensome and the lives of Sami people were subject to numerous restrictions. They were supposed to stick to what was seen as their traditional way of life. They were supposed to herd their reindeer and live in mountainous areas, where they were not in competition with Swedish-speaking farmers. There were periods when they were not permitted to live in rectangular houses, but only in round Sami huts, and they were not allowed to own more than five goats.

Naturally there was constant interchange between Swedish-speaking and Sami-speaking groups. People inter-married and had children together. Many Sami began to increase their reliance on farming as a means of subsistence. But at the end of the nineteenth century and the beginning of the twentieth, Swedish colonialism took on a new form. Ore, forest products and hydroelectric power became more important to the economy. It was in that context that 'racial biology' flared up.

'Racial biology' and 'racial hygiene' (eugenics) constitute a long, dark chapter of the history of ideas in the West. Their worst consequence was the Holocaust during the Second World War, the systematic execution of millions of people – in most cases because they belonged to groups such as Jews and Roma. Another tragic consequence was forced sterilisation, which, in Sweden, affected mainly the people known as travellers – or 'vagrants', as they were called at the time.

One of the fundamental tenets of 'racial biology' was that people belonged to different races, rather as dogs or horses could be of different breeds, and that the intermingling of these races could be disastrous. Miscegenation (racial mixing) would cause the 'racial stock' to deteriorate. One of the sources of inspiration of these murky ideas was genetics, coupled with the new biology, which developed rapidly in the early twentieth century.

The patterns governing heredity that Gregor Mendel discovered in the 1860s had no impact outside Brno in his lifetime, and after his death they were forgotten until the turn of the century. Charles Darwin, on the other hand – Mendel's contemporary – had an immediate breakthrough in the 1860s with his theory of the origin of species. When Mendel's findings were rediscovered, the biologists of the early twentieth century sought to fit the two systems of thought together. And a great deal went wrong in the century's first few decades.

Mendel's genetic laws quickly became an indispensable tool for plant and animal breeding. Crops and animals cross-bred thanks to this new discipline gave better yields, which was important in a society where many people went hungry. In that context, ideas about how genetics should also be used as an instrument of social engineering became increasingly influential – not just to prevent hereditary diseases, but also to overcome poverty and crime and to improve 'the race' in general. It was thought that the people as a whole were becoming increasingly enfeebled, as more and more of them abandoned farming and settled in towns.

These currents of thought were very widely shared across virtually the whole political spectrum: social democrats, conservatives, members of the Farmers' Party (*Bondeförbundet*) and liberals, all of whom had their reasons for taking a positive view of 'racial hygiene'. However, these ambitions for improving 'the race' were increasingly combined with much older ideas about the 'Germanic race', which many in Germany – and Scandinavia too – regarded as culturally superior. This Germanic nationalism had been developing little by little ever since the seventeenth century.

In Sweden, an Institute for Racial Biology was established in Uppsala in 1922. It was one of the world's first completely state-run institutes of 'racial biology', though others were to follow. One of them was the Kaiser Wilhelm Institute of Anthropology in Berlin, a driving force behind the Holocaust that collaborated with Josef Mengele, a German SS officer and physician at Auschwitz concentration camp.

The instigator and first director of Sweden's State Institute for Racial Biology, the medical doctor Herman Lundborg, had an almost manic interest in classifying the people of Sweden – especially the Sami. He worked principally in the contemporary discipline known as 'physical anthropology'. Much of his time was spent measuring crania and comparing family trees from church records. He collected data from over 100,000 individuals, most of them national servicemen, whom he attempted to classify by assigning labels such as 'Germanic type', 'Sami type', 'gypsy' and 'vagrant'. For long periods, he left the institute in Uppsala to its own devices in order to travel around northern Sweden measuring Sami people. He visited villages and markets, measuring people wherever he possibly could. He was accompanied by a photographer who took frontal, rear and side views of his research subjects.

Lundborg's thesis was that the Sami belonged to the so-called 'short-skulled type', while people of Germanic stock supposedly had elongated skulls. However, no matter how many measurements he took, he found no statistical evidence for any such pattern.

He was pensioned off as director in the mid-1930s. A new director was appointed at the State Institute for Racial Biology, and it took a completely new direction, focusing on hereditary conditions and widespread diseases such as tuberculosis. In the 1950s it underwent a change of name, becoming the Institute for Human Genetics, and it later became known as the Institute for Medical Genetics. In other words, the change of name came many years after its actual research had ceased to include cranial measurement and 'racial biology'.

But Lundborg continued to classify people. In 1938 he was awarded an honorary doctorate in Heidelberg, Germany, at the behest of the foremost Nazi race ideologue, Hans F. Günther. He became increasingly anti-Semitic; one of the beliefs he held was that the State Institute for Racial Biology was being undermined through Jewish influence on the press. He became an open supporter of Nazism.

And Lundborg was by no means alone in his racist views on the Sami people. As late as 1947, the Swedish Tourist Board published a book called De svenska fjällapparna ('The Mountain Lapps of Sweden'). The author was Ernst Manker, director of the Nordic Museum's 'Lapp Department'. One of the first few chapters is entitled 'Race and Temperament'. Kjell-Åke Aronsson of the Sami Museum, Ájtte, considers its wording so offensive that he refuses to quote from it. But I want to know what sort of language was used.

I seek out the book myself in the Nordic Museum's library and read, among other things, that the Sami appear to be 'the last vestiges of a distinct tribe descended from a white-yellow aboriginal race' and how they 'are not really very good-looking by Indo-European standards'. It is possible to use 'coaxing and indirect manoeuvres' to 'get these reindeer-herders where you want them, at least to some extent', notes Manker. While Lapps are notably quick-witted and very animated, according to Manker, they are also 'unstable and Bohemian', and they are of a 'protean and malleable' disposition.

Perhaps the most offensive form of words by today's standards is Manker's opinion of the Sami's business sense. He describes their 'limitless cunning and stinginess' and quotes an acquaintance as saying, 'I'd rather make a deal with 10 Jews.'

This book, it is worth remembering, was published two years after the end of the Second World War, at a time when no one, but no one, could deny what had happened during the Holocaust.

There are reasons why relations between some of Sweden's Sami and the Swedish Tourist Board may still be tense. And this background explains why some Sami are suspicious of geneticists with ideas about their early history.

CHAPTER THIRTEEN

Pottery Makes its Appearance

Archaeologists working on excavations at Tågerup in Skåne have identified layers from a period that began 8,500 years ago and ended a full 1,700 years later. The site shows clearly how a first settlement flourished for several centuries at the time of an older culture known as the Kongemose. But then all traces of human life cease for 300 years. The period of extreme cold 8,200 years ago must have dealt a serious blow. However, the coup de grâce for the Kongemose culture seems to have been the flooding of the settlement at about the time that Doggerland disappeared into the depths of the sea. About 300 years later, a new culture known as the Ertebølle made its appearance.

During the earlier period, the Kongemose, life was easier. There was relatively little competition for natural resources. People were few and far between. It was easy to find enough to eat, and people had ample free time. This meant they could focus their full attention on making attractive artefacts from high-quality flint, as well as bone, antlers, wood and plant fibres. Children could spend plenty of time at 'flint school', learning to be skilled and aesthetically aware toolmakers like their elders. That they practised is evident from the half-finished objects they left behind.

Skåne archaeologists interpret the bone remains found in the settlement as showing that the Kongemose people were discriminating and persistent hunters. They selected mainly the youngest and oldest deer, thereby maintaining stocks for future generations. The only buildings at the settlement seem to have been a single longhouse and a few round huts.

By contrast, during the later period corresponding to the Ertebølle culture the settlement was more like a large village with hundreds of inhabitants. There were several large permanent buildings. At the water's edge were elaborate systems of traps made of wooden posts and poles, designed to catch eels and other fish.

These people seem to have indiscriminately hunted anything and everything that was edible. They were not concerned about the age of the deer they shot. Hedgehogs and squirrels were also perfectly acceptable prey. Tools of flint, bone and antler were produced quickly and carelessly. Function was all that mattered now; aesthetics and perfectionism were set aside.

A skeleton reveals that one of their children died with an arrow in his back – perhaps in an internal conflict, perhaps as the victim of a rival group.

Life during the Ertebølle period was clearly harder than it had once been. But then came a significant innovation that may have helped make life easier. People began to use ceramic vessels for cooking.

***

As I described in Chapter 5, the people of Dolní Věstonice made clay figurines, which they laid in the fire to make them explode – perhaps as a serious ritual, perhaps as an entertaining party piece. That was happening a good 30,000 years ago. But making complete earthenware pots for cooking is a different matter – and considerably trickier.

The most ancient ceramic vessels we know of today were discovered in East Asia. At the time of writing, the very oldest examples of ceramics are a few potsherds from Xianrendong cave in China. Having been dated at about 20,000 years, they were made when the last part of the Ice Age was at its very coldest. Other finds from the same location show clearly that the people who lived in Xianrendong cave were hunter-gatherers. Ancient remnants of ceramic pots have also been found at the River Amur in Siberia and in Japan.

The oldest known ceramics in Japan are known as Jomon. The earliest finds date back some 15,000 years. Archaeologists have analysed burnt leftovers found on potsherds. The level of various isotopes in the traces of food shows that Jomon vessels were used chiefly for cooking saltwater fish or marine mammals. Yet these pots were found inland, a long distance from the nearest sea coast. To make an educated guess, the early potters on the Japanese islands may have caught fish such as salmon and eels on their way upstream to spawn. These fish were put in the pots and cooked on a fire to make an early type of salmon or eel soup.

In the past, many archaeologists believed that the art of pottery was first developed when people went over to farming. But that has been proven wrong. The finds from the Xianrendong cave in China date back more than 10,000 years before the first farming at the same location. And it is absolutely clear that the Jomon vessels from Japan were used by people who lived by hunting and fishing.

A more pertinent question today is who first came up with the technology for making earthenware pots. Were these people from Siberia, China or Japan? Did the first farmers in the Middle East learn how to make clay pots from East Asians or Africans, or did they invent the art themselves, independently of anyone else? There is no consensus on this among archaeologists. Some believe the technology must have been invented in one place and have spread from there, while others think pottery may have been developed independently in several places.

In any event, we know that the first traces of earthenware pots in Denmark and Skåne date back some 6,700 years to the Ertebølle hunting culture. Indeed, the world's first example of spiced food is to be found in just such vessels.

Ertebølle vessels were large and had a pointed base. They were ideal for making fish soup for a group of people. The pointed underside prevented food from burning too easily. Chemical analyses of potsherds show that pots in northern Europe were also used to make fish soup. In shards of Ertebølle vessels from Denmark and northern Germany, researchers have also identified microscopic traces of seeds from the *Alliaria petiolata* plant (commonly known as garlic mustard or Jack-by-the-hedge), a relative of mustard that has peppery seeds.

In earlier times, archaeologists equated both pottery and the use of seasoning with farming societies. It was thought that hunter-gatherers lived in a more primitive way, that they wanted only to fill their bellies and lacked time and energy for the finer things in life – such as herbs and spices that provide flavouring but no nourishment. We can now dismiss that view. The world's oldest known seasoned dish is a Nordic fish soup spiced with peppery garlic mustard seeds.

Ertebølle pottery also includes small, shallow bowls that were probably used as lamps. They resemble the stone lamps used by people from the Magdalenian culture when they were creating their cave paintings. Both types of lamp are small bowls designed to hold animal fat; a wick made of plant material would have been dipped in the fat and lit.

So it is clear that hunting peoples in northern Europe were using pottery some 6,700 years ago. The question is how the technology arrived here. Is it possible that the art of pottery came directly from East Asia via hunting peoples in Russia? Some archaeologists think so. An older school of thought is that pottery is more likely to have spread from people who had begun to live on farming in the Middle East.

When I consult Russian archaeologists on the matter, it turns out that there is an element of truth in both views. The very oldest known pottery in eastern Europe is about 8,900 years old and was found at an archaeological site called Rakushechny Yar on the lower reaches of the River Don, a few dozen kilometres from where it flows into the Sea of Azov. The people of Rakushechny Yar kept domestic cattle and sheep even in those early days. Their buildings and stone tools also resemble those of farming peoples in the Middle East, on the other side of the Black Sea.

From the herders of Rakushechny Yar, pottery spread to hunting peoples in other regions of what is now European Russia. This diffusion was slow to begin with. Russian archaeologists believe that clay vessels were initially status symbols used for ritual purposes rather than everyday objects. But ceramic vessels gradually became increasingly common.

The art of making pottery spread further westwards and northwards, and gradually all the way to the Baltic Sea.

By about 7,000 years ago, clay pots had reached the Baltic and Finland, including the Åland Islands. The type of pottery concerned is called Comb Ware, as combs were used to make the typical patterns on the outside.

Ertebølle pottery reached Denmark and southern Sweden some 6,700 years ago. The Ertebølle pottery of the southern Scandinavian hunting peoples shows the influence of Comb Ware from the east. However, it also has technical similarities with a type of pottery used by farmers from more southerly parts of the continent, known as Linear Band Ware. Quite simply, it is a hybrid between the pottery of the eastern hunters and that of the southern farmers. Clay vessels – one of the greatest technological advances ever to reach Sweden – were a mix. Just as the people would be.

CHAPTER FOURTEEN

The Farmers Arrive

Just a few centuries after the earliest pottery found in southern Sweden and Denmark, traces of agriculture also appear.

The question of how that came about has been the most toxic issue ever to dog archaeology in Sweden. In fact, that applies not just to Sweden, but to Europe as a whole. I have been a science journalist for several decades, and in virtually no other area have I heard researchers express such virulent views of intellectual opponents and new findings.

According to the conventional wisdom, which has prevailed since at least the 1970s, hunters reinvented themselves as farmers. The hunters who already peopled southern Sweden gradually began to acquire the various components of agriculture, such as the practice of keeping cattle, sheep and pigs, and the cultivation of wheat and barley. After a while, growing crops and raising livestock became their main means of subsistence. There were no immigrants in the picture – or, if immigration had any bearing on the arrival of agriculture, it was of very marginal significance.

This, for example, is the picture given by the standard work that has dominated Swedish academia for decades, *Arkeologi i Norden* ('Archaeology in the Nordic Region'). The section of this work that covers early agriculture is heavily based on a doctoral thesis from 1984 referring to cattle bones and traces of grain that, it is claimed, can be linked to the older hunting culture and date back approximately 6,200 years.

However, the remains in question were dated solely by the order of the layers in which they were found; that is, by their stratification. This is an extremely unreliable method of dating, as objects can easily shift up or down in the soil. Radiocarbon dating is much simpler and cheaper now than it was in the early 1980s. New measurements show that objects described in the doctoral thesis are considerably more recent than has been claimed. The few pieces of evidence that have played so decisive a role in the writing of Swedish history over the last 30 years are simply not up to standard.

Moreover, the Danish archaeologist Lasse Sørensen has gone through data on early farming in Sweden and Denmark with a fine-tooth comb. He shows clearly that the older view − that hunters learned how to farm on their own − cannot be correct. It is based on misunderstandings and errors. Potsherds bearing impressions of grains have been ascribed to the wrong culture and period, while wild aurochs have been classed as domestic cattle.

In the spring of 2012, a study was published that can be seen as a turning point in an archaeological debate that has persisted for nearly a century − not just in Sweden, but worldwide. A team of Swedish archaeologists and geneticists published detailed DNA analyses of four skeletons, all of which are about 5,000 years old.

Their article appeared in *Science*, one of the world's foremost scientific journals. *Science*'s editorial team considered the news so important that they held a press conference in Uppsala. *Science* staff actually came over from the United States.

During my career in science journalism up to that point, *Science* had only held one other press conference in Sweden. Naturally, I went to Uppsala to hear the researchers, Mattias Jakobsson, Anders Götherström, Jan Storå and Pontus Skoglund, give an account of their findings.

The big news was that they could demonstrate a clear difference between the genetic material of one of the skeletons − an individual from a farming community in Västergötland − and the other three, seal hunters from the island of Gotland. The farmer, who was known as Gök4 after the parish of Gökhem where he had been found, proved to be more closely related to Middle Eastern people than to contemporary hunters from Sweden. It was clear that his origins could largely be traced back to areas of Turkey, Syria and Jordan.

The results tallied exactly with what German researchers' more limited analyses of mitochondrial DNA from ancient skeletons had also shown in the last few years. The Swedish researchers had used the very latest technology to analyse nuclear DNA, which accounts for a far greater proportion of an individual's genetic material. The fact that they were able to demonstrate such a clear difference so far north in Sweden, on the northern periphery of Europe, was a clear indication that agriculture had come to the rest of Europe largely with people from the Middle East.

One of the grand old men of Swedish archaeology, Göran Burenhult, made an appearance at the Uppsala press conference, although he had not been personally involved in the study. He was the editor of the standard work *Arkeologi i Norden* referred to above. Now, after all these years, he was prepared to change his position, he announced to the assembled journalists. The DNA analyses had convinced him that agriculture had actually been introduced into Sweden by immigrants.

***

Personally speaking, I was especially pleased that the oldest farmer in Sweden to undergo DNA analysis came from Gökhem in the province of Västergötland, a strikingly beautiful area near Falköping where the rolling landscape is still as green and thickly wooded as ever. As it happens, my paternal grandmother's parents came from a village very close by.

I already knew that my maternal grandmother, Berta, was descended in the direct maternal line from Europe's earliest hunters. Her – and my – mitochondria belong to group U5. So far, this book has dealt with our foremothers' travels in Europe. Most of the individuals from Europe's earliest hunting population that have been tested have mitochondria belonging to group U. The 9,500-year-old individuals from Søgne in Norway belonged to groups U5 and U4. The 9,200-year-old individual from the island of Stora Karlsö had

U4 mitochondria. All the approximately 8,000–year–old individuals from Motala belonged to groups U5 or U2.

However, the average modern Swede bears hereditary traits inherited from both ancient hunters and ancient farmers – about half from each. Could my paternal grand-mother, Hilda, be a descendant of Europe's first farmers?

I decided to have her mitochondria tested. Hilda is no longer alive, and neither is my father. However, Hilda's mitochondria live on in my four cousins, my two female cousins' children, and my father's brother, Anders.

My uncle agrees to have his DNA analysed. He sends off a sample, and a few weeks later the results arrive. As expected, Hilda's mitochondria belong to group H, one of the haplogroups most typical of Europe's first farming people. The farmer from Gökhem, Gök4, belonged to the same group.

To be more precise, my paternal grandmother's mito-chondria belong to a subgroup called H1g1. On a map, I can clearly see how contemporary members of the group form a broad band stretching diagonally across Europe from the Greek islands in the south-east to Britain in the north-west. 'Very probably that reflects the spread of farming,' comments the German DNA researcher Wolfgang Haak, currently working in Australia, when he helps me interpret the results.

It strikes me that the time has come to pay a visit to my grandmother Hilda's ancient forebears. I plan a journey to the home of some of Europe's very first farmers.

PART TWO

THE FARMERS

Katarina gave birth to Greta.

Greta had a daughter called Johanna.

Johanna was the mother of Anna–Greta, who gave birth to Elin.

Elin was Hilda's mother.

Hilda gave birth to a daughter, Gunnel, and two sons, Göran and Anders.

Göran was Karin's father.

CHAPTER FIFTEEN

Syria

My preference would be to travel to Syria, the location of some of the world's most ancient known farming settlements. But there is a war raging there, and travelling to Syria to write is out of the question. Two Swedish journalist colleagues who entered the country were kidnapped and escaped by the skin of their teeth. Hundreds of thousands of people have been killed. Most of the population are on the run. Those taking refuge in neighbouring countries are numbered in the millions. Many Syrians are coming to Sweden too. This is an exodus whose like has rarely been seen at any time in world history.

In the midst of such human catastrophe, it seems odd to be talking about archaeology. Yet the victims of war also include parts of humanity's most important cultural heritage. News flashes reveal how valuable finds are being smashed to pieces or sold online for huge sums, with the revenue going straight into funding the armed conflict.

Just when I have given up all hope of finding any DNA results from Syria's first farmers, a study actually appears, published in the summer of 2014 by a Spanish research team. The Spaniards have been carrying out excavations in the region for many years, together with their Syrian counterparts. The 2010 season was the last one; since then, the excavations have lain untouched owing to the war. However, some samples of bones and teeth turn out to have reached a laboratory in Barcelona, where researchers have been analysing them for several years.

The excavation sites are called Tell Halula and Tell Ramad. One is on the River Euphrates, near the border with Turkey, while the other is close to the capital, Damascus. Both represent the very earliest stages of agriculture, before people began to use ceramic vessels in this part of the world. The oldest samples come from individuals who lived over 10,000 years ago. Researchers have attempted to analyse the DNA of 63 skeletons, but only 15 of the analyses have borne fruit.

Of the 15 individuals concerned, two have mitochondria belonging to haplogroup H – the same main group common to my paternal grandmother, Hilda, and the farmers from Gökhem in Västergötland. The most common haplogroup among these early Syrian farmers is K. Some have N and HV or one of a few haplogroups that, though extensively represented today on the Arabian Peninsula and in Africa, do not seem to have been spread by the early farmers who migrated to Europe.

Haplogroup H may, of course, have been present in the hunting population of the European Ice Age, particularly in eastern Europe. There are a number of DNA analyses pointing in that direction. However, agriculture is the only way to account for the fact that H is now Europe's most common mitochondrial group. Nearly half of Europe's population have mitochondria belonging to haplogroup H. Most of these people – like my maternal grandmother Hilda – can very probably trace their origins in the maternal line back to the world's first farmers, who lived in parts of Syria.

CHAPTER SIXTEEN

The Boat to Cyprus

I travel to Cyprus. Unlike Syria, this involves a safe and comfortable journey. From the scientific angle, too, Cyprus provides an excellent alternative. The whole island is a kind of outsize laboratory for early agriculture. It enables you to observe developments step by step, as clearly as in a time-lapse film.

On the mainland, where agriculture first developed, it is hard for researchers to tell whether cereals, leguminous plants, goats and sheep are wild varieties or the result of deliberate breeding and selection. It takes centuries of cultivation and breeding to develop the characteristics typical of cultivated varieties.

There is no such problem on Cyprus, which lacks nearly all the wild varieties that complicate the picture. This is because the first Cypriots arrived on the island by boat, with everything they needed for farming on board.

The story is almost like the biblical tale of Noah's ark. In the Book of Genesis, Noah takes refuge from the Flood, the result of rain that persists for 40 days and 40 nights, inundating the whole earth. God instructs him to build a huge boat, an ark, and to fill it with animals: seven males and seven females of the 'clean' sort, and a few of each of the 'unclean' sort.

There are many conceivable reasons for the first Cypriots' decision to leave their native region on the banks of the Euphrates and set off on a long, hazardous sea voyage. Floods resulting from heavy rainfall, just as in the Book of Genesis, are one possibility. At that time, immediately after the Ice Age, the sea level was rising and precipitation was increasing in many places, including northern Syria. While the rain made it easier for the first farmers to grow their crops, there may have been too much water at times.

They may also have chosen to leave their homes because the region was beginning to become overpopulated and short of space. Agriculture almost certainly improved children's survival rate. The population would consequently have grown fast, possibly sparking competition and new conflicts.

Alternatively, the first Cypriots may simply have been driven by curiosity and a spirit of adventure.

At any rate, we know they filled their boats with pigs, dogs, cats, goats, sheep and the occasional cow. They also took with them deer and foxes, which they released with a view to future hunting. It seems plausible to assume that they would have taken young animals, which it would have been easier to fit into the boats. They also took seeds with them to grow wheat and leguminous plants such as peas and chickpeas. However, there is no evidence they took vines with them, as Noah did in the Bible; wine seems to have arrived on Cyprus only a few millennia later.

They must have been well versed in building seaworthy vessels and in navigation. The voyage from northern Syria to the nearest promontory on Cyprus was over 80 kilometres (50 miles). There are researchers who believe they even mastered the art of sailing. However, the predominant theory is that they paddled open dugout canoes.

Interestingly, this voyage in the eastern Mediterranean took place at approximately the same time as when people in Scandinavia were developing their boatbuilding skills, which enabled them to settle all the way up the west coast of Sweden and Norway, as I described earlier. However, the first Cypriots' achievement is even more impressive, as they sailed over the open sea. The voyage must have taken something like 20 hours.

Once they reached Cyprus, the settlers could supplement the seeds they had brought with them with barley and lentils, which grow all over the island. They could pick pistachio nuts, figs, olives and plums from trees growing wild. Their cargo included some tools and ritual objects made of stone, but they were quick to find the best flint on the island.

***

I visit a number of archaeological sites on Cyprus, of which Khirokitia (also spelt Choirokoitia), designated a World Heritage Site by UNESCO in 1998, is the most fascinating.

Like most of the prehistoric settlements I visit, Khirokitia is in a beautiful location, on a south-facing slope and immediately above a river. I can glimpse the sea just a few kilometres away. Khirokitia was the site of a fairly large, densely built-up village, full of small, circular houses and surrounded by a high, thick stone wall. The round dwellings, intensely white in the sunlight, would have been visible from a long distance. Some were probably decorated on the outside with patterns executed in red pigment.

Archaeologists and local craftsmen have reconstructed a few houses and sections of walls. They have used the traditional techniques employed by Cypriots well into the twentieth century: outside walls and foundations of limestone and mortar, inside walls of sun-dried bricks mixed with chopped reeds, and flat roofs covered in turf. These building methods have remained essentially unchanged for 10,000 years, except that the houses became square after a few millennia. Locals speak with enthusiasm of how well the traditional houses are suited to the climate; turf roofs and sun-dried clay provide good protection against the heat of summer.

A neatly walled limestone staircase leads up to a gap in the wall. This is the way into the village, which would have enabled the villagers to check who was entering or leaving. However, it is unclear why they would have needed to do so, as there is no evidence of any violent events at that time. Cyprus must have been extremely thinly populated. An archaeologist points out that the wall encircles the western side of the settlement. It may equally well have been designed to shield the village against the strongest winds, rather than human enemies. Alternatively, it may have been intended to keep sheep, goats and pigs out of the houses.

My eyes half closed, I try to picture the scene when the village was full of people. There are fires burning in open courtyards between the houses. Women are seated in groups in these courtyards, working with large stone mortars. I notice one particular middle-aged woman who is pounding her mortar with circular movements. She is spare and slight, with delicate features, brown eyes and dark brown hair. In fact, she resembles my paternal grandmother, Hilda. A cat lies purring at her side.

There are big, rangy cats on the prowl everywhere. People seem to treat them with a respect bordering on reverence. Though there are dogs too, they are quite small. And children abound.

From the field beyond the wall, people can be heard calling out or talking to one another. Are my ears deceiving me, or is there something reminiscent of modern Basque in their language? The people in the field are bent double. Bearing large, crescent-shaped sickles, they are gathering in sheaves of grain. Up in the mountains I spot a flock of sheep, closely guarded by a few shepherds and their small dogs.

Although visitors are forbidden from entering the reconstructed houses in Khirokitia, you can peep in through the door openings. The windows are open apertures. There is beauty in the fall of the light in the round, limewashed rooms. One of the floors contains a grave. In the grave lies a skeleton in a crouched position, weighed down by a large, heavy stone on its ribcage. The living and the dead occupied the small round houses together. The purpose of the heavy stone on the skeleton's chest, according to the archaeologists' theories, was to prevent the dead from rising.

I ask the archaeologist Carole McCartney, who has lived and worked on Cyprus for many years, why these people chose to make their houses round. 'Why not?' she replies. 'Why do we make ours square today?'

That is a reasonable question, of course. First of all, I reflect that circular houses built in stone and brick are simply a natural extension of circular tents. Nomadic people have always made such tents: a few poles in a ring, leaning in towards the apex, covered with animal skins, foliage, bark or whatever material is available. The forebears of the first farmers must have had round tents too, which is why they continued to build round houses of stone – simply in line with tradition.

But later I read an article by two researchers working in Jordan, Ian Kuijt and Bill Finlayson, and the penny drops. I am reminded that we still have round buildings, including in Sweden. They are widespread in rural areas. Farmers use round silos for grain storage. The round buildings of the early farming cultures may also have been a type of silo.

Many of the buildings in Khirokitia and other early farming settlements on Cyprus contain the remnants of two stone plinths. Archaeologists think these supported a wooden platform. Such platforms would have served to store grain, protecting it from damp, mould and mice.

Carole McCartney thinks the earliest stone buildings had an important function as storehouses. In her view, we cannot be at all sure that people lived there all year round. At the height of summer, some people may have gone down to the sea to fish, where they may have lived in simple fishing camps of which no trace has been preserved.

She also points out that people are unlikely to have continued living in a house where they had just buried a dead family member. They may have buried the body under the floor and then left the house for at least a year, until only the skeleton was left and the air was clear again.

Ian Kuijt and Bill Finlayson, the researchers working in Jordan, take the argument a step further. They think the circular buildings served primarily as granaries – and that they were a decisive factor when people switched from hunting to farming.

***

Kuijt and Finlayson have examined several archaeological sites in Jordan, the most important of which, Dhra', lies in the southern Jordan Valley, near the Dead Sea. It was in use at precisely the period when people in the Middle East were beginning to make the transition from hunting and gathering wild plants as their sole means of subsistence to farming the land.

Archaeologists at Dhra' have found a number of small buildings, all of which are circular or oval in shape. Built of stone and mudbrick, they are partly sunken into the ground. Some appear to have been primarily dwellings. However, there are also some round structures that seem to have been used as granaries. The oldest of these is about 11,300 years old.

Just as in Khirokitia, built over a millennium later, these granaries had raised wooden floors supported by plinths. In many cases the floor would have sloped — awkward in a dwelling, but practical for grain storage. Quantities of barley and oat seeds have been found next to the round buildings, together with stone mortars and storage vessels made of woven plant fibres daubed with clay to make them watertight.

Similar remains of granaries have been found at a number of other sites along the River Jordan, in northern Syria and south-eastern Turkey.

There are no definite signs that the people living in Dhra' actually cultivated grain themselves. They were probably gatherers, just as previous generations had been. They would have gone in search of barley and oats growing wild on the mountainsides, cut the stalks with flint tools and carried the sheaves home to their camp. Everyone there would have shared the food, as hunter-gatherer peoples had always done.

The difference was that the people of Dhra' had begun to develop methods for storing grain for extended periods of time, which meant they were no longer obliged to be so mobile. Increasingly, they could settle down. Young children, freed from the strenuous business of continual migration, were more likely to survive. The population grew, and the very first signs of private wealth and a class-based society began to emerge.

According to Kuijt and Finlayson, grain-storage technology was *the* most critical factor on the way to an agricultural society. In their view, the first granaries were the threshold to civilisation.

And there were sentinels guarding that threshold.

***

Two things happened when people began to accumulate large stocks of grain. Firstly, the grain attracted mice. Secondly, the mice attracted cats. People had every reason to be delighted when cats began to eat the mice that were rife in their grain stores. They no doubt did all they could to encourage these cats to stay in their villages, by feeding them, playing with the kittens and even petting the full-grown cats – if the adult cats would let people stroke them.

Today, cats are the world's most common domestic animal. The domestic cat is a close relative of its wild counterpart, *Felis silvestris*. DNA studies suggest that all the domestic cats in the world are descended from a subspecies of *Felis silvestris* living in the Middle East.

It was more difficult than usual to draw up a genealogical table for the world's domestic cats, as tame cats often run off and interbreed with local wild cats. But in 2007 an extensive study was published that compared the DNA of nearly 1,000 wild and domestic cats from all over the world. The results show that the domestic cat's closest wild relatives today live in remote desert areas in the Middle East.

So cats come from the desert. Even today, many domestic cats have retained their natural camouflage colouring; they are still sandy and grey in colour, with stripes on their backs.

For a long time, archaeologists' oldest evidence of domestic cats came from the Egypt of the Pharaohs. Back in the 1940s, a mural painting nearly 4,000 years old of a cat attacking a rat was found in the grave of the provincial governor Baket III. There is also an alabaster vessel of nearly equal antiquity which skillfully depicts a cat, with eyes of rock crystal and copper.

In the necropolis of Abydos, 17 complete skeletons of cats have been found that were apparently sacrificed nearly 4,000 years ago. In recent years, Belgian archaeologists have found even more ancient burial places in Egypt containing the skeletons of cats that appear to have been tame. They were buried nearly 6,000 years ago.

During the millennia that followed, a veritable cult of the cat emerged in Egypt. A number of Egyptian goddesses were believed to take feline form. The best-known is Bastet, who symbolised motherhood and, according to some sources, fertility and sexuality as well.

Temple employees began to rear cats on a truly industrial scale. Visitors to temples could pay for a cat to be killed and mummified. The visitor could then offer up the mummified cat to the gods. Underneath the temples were long subterranean passages whose walls were lined with small niches in which these votive offerings were placed. One excavation of the temple at Bubastis revealed several thousand cat mummies.

Herodotus, the Greek historian, described about 2,450 years ago how people in the Nile Valley worshipped Bastet by holding huge processions, bathing in the Nile, drinking copious quantities of wine and sacrificing cats. That Christianity has often associated cats with the Devil may not be so strange. The early Christians, after all, were confronted by a widespread cult of the cat.

Finds made by French archaeologists in Cyprus have extended the history of the domestic cat by several millennia. When the first settlers came to Cyprus by boat, their cargo included both dogs and cats; there were no wild cats among the island's indigenous fauna.

The oldest remains of cats' bones in Cyprus date back 10,600 years. But the most significant find, from a village called Shillourokambos, is about 9,500 years old. A human being and a cat were each assigned their personal burial place, just 40 centimetres (16 inches) apart.

The human, of indeterminate sex, was around 30 or perhaps a little older. He or she was interred with unusually rich grave goods: a seashell, a stone pendant, a fragment of ochre and several flint tools of different kinds. A further

24 seashells were buried in a small pit right next to the grave.

While the cat was not accompanied by any gifts for the afterlife, it was clearly placed in a newly dug grave of exactly the right size and immediately covered with earth. It was a large, rangy animal – as big as wild tomcats of the subspecies living in the Middle East. About eight months old at the time of its death, it may have been killed to accompany the dead person in the neighbouring grave.

The French researchers believe the two graves demonstrate strong bonds between people and cats. Cats did not benefit people just in practical ways – by hunting mice – but also held a spiritual significance. This view of cats as possessing a spiritual status is also supported by a number of figurines made of stone and clay found in the settlements of the earliest farmers in Cyprus, Israel, Turkey and Syria. A stone cat figurine predating the cat's grave has also been found in Shillourokambos.

One can speculate as to whether the person in the grave was a shaman, a priest or a priestess, with the cat as totemic symbol and closest assistant. The two may have been associated with a combined granary and temple.

In south-eastern Turkey and northern Syria, archaeologists have found a number of large, round stone buildings that stood in the middle of villages. They were considerably larger than the houses there – up to 10 metres (33 feet) in diameter. Just as in Dhra', Jordan, some of them have been interpreted as being granaries. Hiding places containing objects of value such as flint tools, shells and greenstone beads have been found in the walls and the floor. These buildings also appear to have been places where people congregated for ritual purposes. In other words, they were a kind of early temple.

The most ancient settlement in Cyprus that we know of today is called Klimonas. People began to settle there at least 10,800 years ago. Klimonas also had a large round building in the middle of the village. Its inhabitants used flint tools whose appearance shows that their makers were probably from northern Syria. They also brought both dogs and cats with them, as shown by finds of bones and jaws.

***

It is quite conceivable that the cat was the first to take the initiative when humans and cats started cohabiting. To that extent, the story of the cat resembles that of the dog; many researchers think it was dogs that domesticated us, not vice versa. But to an even greater extent than dogs, cats live in human society on their own terms. They differ in several significant ways from dogs, as well as from most other domestic animals.

Nearly all other domestic animals originally lived in herds, packs or flocks. This is why they readily know their place in a hierarchy, such as the pecking order in a henhouse, a barn or a family. Wild cats, on the other hand, live alone; they are more solitary animals that get together chiefly for mating purposes. Eleven thousand years of cohabitation with humans have only partially mitigated that trait.

Dogs, and many other domestic animals, have undergone genetic changes making them more childlike. As a result of this arrested development, they have become more like puppies, both psychologically and physically – more rounded, plumper, more good-humoured and playful, less serious and aggressive. Cats, on the other hand, have undergone relatively little genetic change since the times when they lived alone in Middle Eastern deserts. They have remained adult, retaining their wild relatives' traits far more than other domestic animals.

In fact, all this begs the question of whether cats can really be regarded as domestic animals at all. The term used by many researchers is 'semi-domesticated', 'domesticated' being the term for animals or plants bred so as to be of service to humans.

In November 2014, an international research team published a large-scale comparison between the DNA of wild and domestic cats. They identified 13 genes showing marked differences. These genes have clearly changed in cats descended from those cohabiting with humans for the last 11,000 years. Most of the genes identified express themselves in the brain. They have an impact on fear and on the brain's reward system. Domestic cats also seem to have marginally better night vision. This may reflect an adaptation going back to the time when they used to hunt mice in dark granaries.

The researchers concluded that bold cats which enjoyed being stroked and petted had a better chance of survival around people. Initially, wild cats were attracted to human villages by food, not least the mice proliferating in the granaries. Even then, the most daring cats had an advantage, as they were not frightened off by the presence of humans. Cats that actually allowed people to scratch them behind the ears got on even better. They had a better chance of surviving and having kittens that would also be well cared for. And the more accommodating and ready to learn the kittens were, the more likely they were to live longer and have kittens of their own.

So that's the way it is. The cat's journey towards full domestication continues even today, and many are only halfway there.

CHAPTER SEVENTEEN

The First Beer

Humans invented agriculture in different parts of the world, each group presumably independently of the others. The indigenous peoples of South and Central America cultivated various types of squash as early as 10,000 years ago. People in South-east Asia domesticated pigs at least 8,500 years ago. But the world's oldest traces of agriculture are to be found in the Middle East.

Searching the scientific literature for the cradle of early agriculture, I discovered a curious fact. To my surprise, I ended up just where this book begins – in the region where anatomically modern humans interbred with Neanderthals.

Sex with Neanderthals may appear to be an anomaly, a mere detail of our prehistoric past. Yet that was how our life outside Africa began. A tiny amount of genetic material from the Neanderthals probably made it easier to live in the cool climate of Europe. The traits passed on by the Neanderthals enabled our immune defence, digestion, skin and hair to adjust, helping us to cope with a new and harsher environment.

We know we must have received this Neanderthal input somewhere along the way while modern humans were migrating from Africa to Asia and Europe. However, all that DNA researchers can say – rather vaguely – is that this happened about 54,000 years ago, 'somewhere in the Middle East'.

This prompted me to turn to Ofer Bar-Yosef from Israel – a veteran of Middle Eastern archaeology – and ask him to suggest a more specific location. He proposed Galilee as the most reasonable possibility. The region was rich in resources that would have supported both a Neanderthal population and the newly arrived modern humans. There was drinking water, plenty of gazelles and other animals to hunt, and fruit, nuts, herbs and nutritious grasses to round off people's diet.

After our conversation, the discovery of a skull confirmed that modern humans really did live there 55,000 years ago, a mere 40 kilometres (25 miles) away from the Neanderthals.

Having grown up in a Christian country, there are few places I have heard more about than Galilee. My first two years at school were devoted almost exclusively to the deeds of Christ in the region. The Annunciation, when the Angel Gabriel told Mary that she would conceive through the Holy Spirit, was set in the Galilean town of Nazareth. It was on the shores of the Sea of Galilee that Jesus called his first disciples to him. There, too, they witnessed the miracle of an extraordinary catch of fish in the fishermen's nets.

A fishermen's camp on the shores of the Sea of Galilee has also provided the earliest evidence of people living on wheat, barley and oats – the cereals that laid the foundations for European agriculture and civilisation. The archaeological site at Ohalo is so well preserved that it constitutes a small miracle in itself. This fishing settlement, inhabited some 23,000 years ago, comprised a number of oval huts built simply and rapidly from branches, foliage and grass, with a few fireplaces in between.

One day, the hamlet was struck by a large fire that razed the huts. Almost immediately after that, the burnt remains were engulfed by the rising waters of the lake. Naturally, these two events were catastrophic for the hut-dwellers. But 23,000 years later, the fire and the rapid flooding proved to be an almost incredible stroke of luck. Fire and clay have preserved the contents of the settlement, providing archaeologists with almost ideal conditions for reconstructing life as it was lived at Ohalo.

As a result, we know that the settlement's inhabitants lived largely on fish from the lake, but also hunted gazelles, deer and a wide variety of birds. They gathered pistachio nuts, almonds and olives from the trees, as well as wild grapes and a wide range of herbs. In total, archaeologists have identified the remains of over 150 species in the settlement.

The most significant find of all is a large, flat stone held down by smaller stones. This is the world's earliest known millstone, on which the people of Ohalo ground the seeds of starchy grasses by moving a smaller stone back and forth.

Israeli researchers have examined microscopic traces of starch from the stone, enabling them to establish that it was used for grinding emmer wheat, barley and oats. They have also identified traces of these grasses on the ground around the millstone, in the huts and around the fireplaces.

Although archaeologists have found small, simple mortars from even earlier times in Italy, Russia and the Czech Republic, the Ohalo millstone shows that milled grains accounted for a far larger share of the Ohalo people's diet.

The people of Ohalo lived from fishing and as hunter-gatherers. They cannot by any stretch of the imagination be described as farmers. Yet here we see the first large-scale use of the crops that would eventually change humankind's living conditions in so radical a way.

The fact that Ohalo has remained so well preserved is the result of extraordinary circumstances. Wild grains may have been part of people's staple diet over a wider area, without any remains having been preserved for posterity. Yet just as we can say that we must have acquired our Neanderthal genes in the Middle East, we can be quite certain that agriculture made its first appearance in that region.

A number of researchers claim that this was a drawn-out and geographically diffuse process involving the whole region known as the Fertile Crescent. The region stretches from the Jordan Valley to northern Syria and south-eastern Turkey, down the valleys of the Euphrates and the Tigris, and towards the Zagros Mountains of Iran.

Other researchers believe that agriculture developed in a far more concentrated way, in terms of both time and place. They contend that the whole package emerged in the course of just a few centuries, in the border regions of Syria and Turkey. In my view, DNA and other clues now point more in this direction.

***

The first clear signs of agriculture in the Middle East appear at the end of the Ice Age, some 11,500 years ago. This may of course be a coincidence. Some archaeologists strongly disapprove of attempts to explain human behaviour in terms of climate change; they prefer to emphasise internal psychological and social motives.

However, many of those studying the issue perceive a close link between the altered climate and the first signs of agriculture. That link emerges very clearly with the hunting people known as the Natufians, who lived in Jordan, Israel, Palestine, Lebanon and Syria for millennia before any trace of agriculture appeared.

The Natufians seem to have been at least partially sedentary, and their villages contain vestiges of small, round houses constructed on brick-built foundations. However, the Natufians' houses were a great deal simpler than those we see in Cyprus and in other early farming settlements. They were more like sophisticated huts, with simple stone foundations, but without brick walls. The roofs were conical structures made of branches, foliage and grass, rather than the flat timber roofs covered in peat that emerged a few thousand years later.

Natufian culture first blossomed between 14,500 and 12,700 years before our time, which corresponds exactly to the first warm period after the Ice Age. This was the time when the first known dog was buried at Bonn-Oberkassel together with two of my forebears from group U5b1, as I described earlier. And it was at exactly the same time that warmth drove the reindeer northwards, followed by their human hunters. People visited the region now known as Sweden for the first time, and life thrived in Doggerland, which now lies beneath the sea.

The world's first farmers had exactly the same origins as the hunters who were the first to people Europe. They were descended from the same small group of people who had left Africa and interbred with the Neanderthals some 54,000 years ago. But while my maternal grandmother's female forebears continued their journey to Siberia and Europe, my paternal grandmother's foremothers – who were to become farmers – stopped somewhere in the Caucasus or the Middle

East. The two groups lived separate lives for over 30,000 years, after which they met again.

Before the farmers became farmers, they hunted gazelles and picked almonds, figs and pistachio nuts. They gathered grain – of slightly different types, depending on where they happened to live – using sharp flint tools. Stone mortars 14,000 years old, which the Natufians used to crush grain and other foodstuffs, have been found in the Jordan Valley. They look just like the marble mortar on my kitchen shelves at home.

Twelve thousand seven hundred years ago, the Natufians suffered major setbacks. The Ice Age returned with a vengeance for a thousand years as the period of cold called the Younger Dryas. The climate became both much colder and much drier. The cold was so severe in northern Europe that people were forced to leave the region that is now Sweden. In the Middle East, it was not so much the cold that was the problem as the dry climate. The Natufians' main prey, gazelles, dwindled – as did nuts, fruit and other vegetable products.

The Israeli archaeologist Ofer Bar-Yosef has studied this period for over half a century. He believes the Natufians reacted in three ways when conditions grew harsher during the Younger Dryas. Some migrated, leaving the regions where life had become too hard. Others actually became more sedentary than they had been in the past and began to build walls and defensive structures to keep out rival groups. And some started to cultivate grain and legumes, which they had previously gathered in the wild.

Eleven thousand six hundred years ago, the Ice Age came to a definite end. Precipitation increased considerably in parts of the Middle East, particularly in northern areas such as northern Syria and south-eastern Turkey. The winters became mild and rainy, and the climate became considerably more stable. This was a boon to those who had begun to experiment with cultivating grain. The new climate provided an ideal environment for an early farmer.

It was at just this time that a completely new type of jewellery appeared – perforated greenstone beads. Ofer Bar-Yosef has suggested that these green beads may indicate a new kind of cult; they may represent and bear homage to cultivation and budding green plants. Though that may be a trifle far-fetched and hard to prove scientifically, it's a poetic thought.

Evidence of early barley growing can also be found in the most south-easterly extremes of the Fertile Crescent in Jordan and the Zagros Mountains of Iran. However, the archaeological sites that are particularly significant in terms of the transitional phase lie in northern Syria and south-eastern Turkey. Researchers are sure there were networks in these regions that extended over very long distances: over 12,000 years ago they stretched from south-eastern Turkey to the Euphrates Valley in Syria, and a few thousand years later, they extended as far as Jericho in the Jordan Valley and Cyprus. The researchers concerned can identify the provenance of the obsidian – a type of volcanic rock – found in the various settlements. Obsidian is a hard, glassy stone that was sometimes used to make tools. Thanks to isotopic analysis, individual pieces of obsidian can be matched with the volcanoes they came from.

A Franco-Polish research team led by the botanist George Wilcox has published the results of some of the most recent investigations of the Tell Qaramel settlement in northern Syria. This was inhabited when the cold period known as the Younger Dryas came to an end and the subsequent warm period began.

The people of Tell Qaramel ate almonds, pistachios and a type of cherry. There are many traces of einkorn wheat, lentils and peas. Researchers have also found a number of species that they interpret as being weeds. Such plants flourish on open land, which is very probably the result of hoeing by Tell Qaramel's inhabitants.

Interestingly, the meticulous methods of various research teams have revealed large numbers of mouse droppings. Mice clearly flourished in a setting where people accumulated extensive grain stores. They must have been a real nuisance. They were a threat to people's food and – not least – a threat to beer.

***

It was 1997. I had been working for just under a year as science editor at *Dagens Nyheter* when *Science* published a groundbreaking study on the origins of wheat. To this day I still feel a certain pride in the science page we produced that Sunday. It was unusually decorative, with an attractive watercolour of a wild wheat plant by Sweden's top botanic artist, Bo Mossberg. 'The wild origins of wheat' was the headline. It was one of the studies from 1997 that opened my eyes to DNA technology's potential for unravelling the early history of humankind.

A team of German and Italian researchers had taken DNA samples of einkorn wheat, one of the two types of wheat grown by the first farmers. They tested numerous samples of both cultivated and wild varieties from a region extending from Turkey all the way to Iran. When they compared cultivated varieties with wild wheat, everything pointed towards a small area in south-eastern Turkey, lying on the western slopes of Mount Karacadağ, a few dozen kilometres west of Diyarbakır.

The study's first author was the biologist Manfred Heun, German by birth but employed by the Norwegian Agricultural University in Ås. Heun is among the scientists who believe that the first plants to be cultivated must have developed very rapidly, possibly over as short a period as 30 years. He also argues that this development began at one location. Over the years, Heun's hypothesis has been heavily criticised, notably by French researchers who believe the transition took much longer and that individual hunting populations over a large area – extending from southern Jordan and Israel to Iran – experimented with cultivation. But later, more sophisticated DNA analyses support Heun's hypothesis.

American and Chinese researchers have also studied emmer wheat, the other type of wheat grown by the first farmers. Its origins have also been traced to the area around Diyarbakır in south-eastern Turkey.

The early farmers' 'starter package' is generally thought to include seven crops. Three of these – einkorn wheat, emmer wheat and barley – were cereals. Three were legumes: lentils, peas and chickpeas. The seventh crop was flax, grown both for its fibres and for its seeds, which are rich in linseed oil. Fig trees also appeared very early on, as did vetch, another leguminous plant.

A notable feature of chickpeas is that the subspecies first cultivated grew in a more restricted area than the other species in the 'starter package'. It was found only in a small area. This implies that cultivation must have begun in the areas around the Syrian–Turkish border.

Current findings thus suggest that early cultivation of at least emmer wheat, einkorn wheat and chickpeas began in the same area. The seed then spread to fertile areas throughout the Middle East. In many cases, these crops were later mixed with wild local species.

The area identified by DNA studies is very close to the Euphrates and the Tigris. It is only a few dozen kilometres away from the Turkish archaeological site of Nevalı Çori, where archaeologists have found early traces of cultivated grain. And you do not have to follow the Euphrates very far before reaching sites in Syria that have even earlier traces of agriculture, such as Mureybit, Abu Hureyra, Jerf el-Ahmar and Dja'de.

But what has struck Manfred Heun most in recent years is these sites' proximity to Göbekli Tepe in Turkey. He and many other archaeologists believe that the excavation at Göbekli Tepe is one of the most important events in archaeology for decades.

Göbekli Tepe was a place of worship high up on a mountain peak, which could be seen from a long distance. Bands of hunters met here to hold celebrations. This began about 11,600 years ago. Layers of crushed bones from wild animals show that the visitors feasted on aurochs, gazelles and wild asses, and that they crushed the bones to extract the rich bone marrow.

These hunters devoted a huge amount of effort to building their place of worship. They constructed large, round limestone buildings, whose inner walls were lined with benches. In the walls and inside the rooms they erected enormous T-shaped stones. They decorated these stone pillars with carved figures of people and animals including serpents, scorpions, wild boar, birds, aurochs, bears and foxes. They also used small drinking vessels of stone bearing similar motifs.

The remains of other sites with similar round buildings and huge pillars have been found in nearby areas. But Göbekli Tepe was the largest. Unlike the other sites, Göbekli Tepe had no dwellings, nor was there any nearby spring. No one can have lived there; this was a place where people met on special occasions. Scaling the mountain was hard, and it was an impractical business. However, the advantage of the site's elevated location was that it could be seen for miles around.

Archaeologists have also found the same type of small, skilfully fashioned stone drinking vessels, similarly decorated with snakes, scorpions, birds, wild boar, aurochs, bears and foxes, at several other local sites. They have been found both in Turkey, at archaeological sites such as Çayönü and Nevalı Çori, and in Syria, at places including Jerf el-Ahmar, Tell Abr and Tell Qaramel. Tools and small stone plaques decorated in a similar style have also been found on both sides of the border.

And the people who gathered at Göbekli Tepe were in contact with people even further away. This can be seen from finds of the volcanic rock obsidian, which came from a number of different places. Some stones come from the region of Cappadocia in central Turkey, about 500 kilometres (310 miles) to the west. Others are from the area around Lake Van, over 200 kilometres (125 miles) to the east. And some are from volcanoes around 500 kilometres (310 miles) to the north.

This spectacular place of worship with its round temples and gigantic stone pillars thus appears to have been the nerve centre of a far-flung network. The German archaeologists concerned believe that Göbekli Tepe was at the heart of an emerging cult.

The largest T-shaped stones weigh up to 50 tonnes (110,000 pounds) – as much as a lorry loaded with timber. It may seem inconceivable that Stone Age hunters could have dragged such huge blocks of stone from cliffs hundreds of metres away and then raised them. The question is: what motivated them? The German archaeologists have a hypothesis. They are convinced that the workers fortified themselves with beer.

There is sound evidence from written sources that the slaves who built the Egyptian pyramids several thousand years later were given special rations of beer. The evidence from Göbekli Tepe is more uncertain at the moment. Apart from the finely worked small drinking vessels, the archaeologists have found about 10 large limestone vessels that they believe were used to add malt to grain and brew beer. The vessels, which resemble casks or bathtubs made of limestone, are set in the middle of the rooms. The most capacious is large enough to hold 240 litres (420 pints). Chemists have looked for traces of a salt called oxalate that is formed during brewing. They have found a few such traces. However, it has not proven possible to reproduce the results, and they need to be confirmed by further, more sensitive analyses.

Manfred Heun and his colleagues may well be correct in hypothesising that beer, brewed from einkorn wheat and barley, drove not only Göbekli Tepe, but all early agriculture. Personally, I think it seems very reasonable to see alcohol as a strong motive force. After all, extraordinary circumstances must have been needed for a handful of pioneers to abandon hunting as a way of life, when it had kept humankind alive for hundreds of thousands of years. Providing the ingredients for beer and wine in order to hold great ritual celebrations may very well have been the necessary motive.

We can only speculate about the possible content of these rites. We know nothing of these people's songs, dances and myths, even though the scorpions, serpents, birds and wild quadrupeds depicted on the stone drinking vessels and pillars provide a few clues. However, we do know quite a lot about the outer forms of the cult today, thanks to Göbekli Tepe and other nearby sites. The people who met here apparently enjoyed grilled pork. They probably drank beer or other alcoholic beverages made by blending honey, grapes and grain.

Somewhere and somehow, alcohol entered our lives as a source of pleasure and a way of enhancing celebrations. But at the same time, the curses of alcohol – abuse, violence and addiction – made their appearance. Ten thousand years later, both aspects continue to play a very prominent role in life in Europe.

***

On the European mainland, it is hard to distinguish between domesticated and wild pigs or between cultivated and wild wheat. On Cyprus, however, the picture of early agriculture emerges far more clearly. Here we can watch the film frame by frame. We can definitely say when and how the first domestic animals arrived on the island. The story began with the pigs that turned up a good 12,000 years ago. They came by boat.

During the Ice Age there were no people living on Cyprus. The largest animals among the island's original fauna were dwarf elephants and dwarf hippos. Both species had shrunk to miniature versions of themselves, the result of a biological mechanism that often affects species on isolated islands.

The hippos and elephants died out towards the end of the Ice Age. Human hunters were very probably involved. Inside a British military base, amateur archaeologists have found the remains of elephant bones immediately under a steep cliff. Their conclusion is that the people of the time drove the elephants over the cliffs – an effective hunting technique.

These remains, which are some 12,500 years old, come from an archaeological site called Aetokremnos. The site was a storage space protected by a cliff, which, at the end of the

Ice Age, lay a few hundred metres from the shores of the Mediterranean. Archaeologists have found a number of flint tools and large quantities of food waste at Aetokremnos: seashells, fish bones and bones from birds.

They have also found 18 teeth and pieces of bone from small pigs. Their interpretation is that hunters from the mainland sometimes went to Cyprus by boat to hunt birds. They ended up taking piglets with them, which they left to return to the wild on the island so as to supplement the meat they obtained from birds. Since the dwarf elephants and dwarf hippos – which were about the same size and had roughly the same dietary habits as pigs – had recently died out, there was an ideal ecological niche for the pigs to occupy.

A number of researchers have already suggested that the pig farmers' ancestors in south-east Turkey kept herds of half-wild pigs, just as today's Sami in northern Scandinavia herd reindeer. The animals could roam about quite freely with a minimum of control by human beings. People knew where they were, but basically left them to fend for themselves. At appropriate times, when the people wanted some meat, they would herd the animals together and shoot them with bow and arrow. The finds of pigs' bones on Cyprus supports the theory that there was a half-wild transitional stage before pigs became fully domesticated.

The first people to settle permanently on Cyprus were early farmers. They came to the island by boat, bringing emmer wheat, cats, dogs and their typical greenstone beads. They built round buildings like their relatives on the mainland, including a large, round meeting house in the middle of the village with stone benches lining the walls. The bricks to build the walls of their houses were made by adding large quantities of chaff from emmer wheat and barley.

They may have found the barley locally, as it grows wild on Cyprus. But they must have brought emmer wheat seed with them from the mainland. They used small, oval millstones to grind the grain. The flint in their sickles shows the typical shiny surfaces that come from frequent use to cut cereal stalks, which contain silica. There can be no doubt that the inhabitants of the oldest settlement, Klimonas, grew some of their food. To obtain meat, they hunted the feral pigs that their ancestors had released on the island several hundred years previously.

In the mountains 30 kilometres (20 miles) north of Klimonas lay another village, Asprokremnos, which was inhabited almost as early and had the same type of flint tools. Its inhabitants seem to have made fewer efforts to grow cereals; they apparently focused even more on hunting pigs. Asprokremnos may have been a hunting camp used by the same group of people who lived in Klimonas. Alternatively, it may have been inhabited by another group who had not taken up farming to the same extent.

The next animals to arrive on Cyprus were goats and cattle – and, unfortunately, mice, which the boatmen from the mainland must have taken with them involuntarily in their cargoes of grain. The goats were released and left to their own devices so they could be hunted in due course like the pigs. There are traces of goats, cattle and mice from two settlements that both came into being about 10,400 years ago: Shillourokambos, where the first cat burial was discovered, and Mylouthkia, the site of the world's oldest known wells.

I visit Mylouthkia, which lies on the west coast of Cyprus, together with the archaeologist Carole McCartney. She was born in the US, but has lived and worked in Cyprus for many years. The house where she lives with her Cypriot husband and her children is just a few kilometres from Mylouthkia. The location is well chosen; streams of pure water run through the limestone hills, and to the west you can watch the sun sink into the shimmering sea.

To our disappointment, the world's oldest wells turn out to be in a poor state. Builders have broken them while widening a road up to a new hotel. I can still see from the remains how meticulously fashioned they once were: up to 11 metres (36 feet) deep, lined with stone, and with a little set of steps made of jutting stones on the inside.

In one of Mylouthkia's wells, archaeologists have found the bones of pigs, goats and mice, and even the skeleton of a young woman.

The layers representing the first stages of settlement in Mylouthkia and Shillourokambos are largely dominated by the feral pigs. While cattle and goats do occur, they are rare. In the following centuries, goats become much more common, and there is a new addition – sheep.

About 10,000 years ago, new boatloads of cargo seem to have arrived from the mainland, including a new generation of pigs. These new animals seem to have been better adapted to a life alongside people, according to archaeologists' interpretations of their size and the shape of their legs and teeth. It looks as though the hunting of wild pigs came to an end. Instead, the villagers began to keep domesticated pigs. The island's wildlife was enriched by two new species from the mainland: fallow deer and foxes. The deer were hunted for their meat. The skins of both deer and foxes would certainly also have been used to make clothes.

The relative numbers of ewes and rams and of young and old animals suggest that the sheep were initially kept for both milk and meat. After a few centuries, however – about 9,500 years ago – the first generation of Cypriot sheep seems to have died out. Presumably they became too inbred in their isolated island existence, as the poor state of their teeth and skeletons suggests. After this, new boatloads of domestic sheep from the mainland arrived on Cyprus. They looked slightly different and seem to have been kept essentially for their meat.

There are clear signs that people took a stronger grip on the flocks of feral goats at the same time. They killed the male kids, but kept the nanny goats for milking. It looks as if the goats replaced the sheep as a source of milk when the latter began to be used more for their meat and wool. Cattle appeared from time to time, but they were rare. They are more difficult to breed, and they must have been too large and ungainly to get into boats easily. Moreover, they are not particularly well suited to Cyprus's dry climate.

***

Nine thousand five hundred years ago, the island's inhabitants had settled down into an annual rhythm that seems oddly familiar. It is not hugely different from the way people lived in rural areas well into the twentieth century. In the autumn they hunted fallow deer in the mountains. In winter they slaughtered their pigs. In spring and in the early summer, they slaughtered their lambs and harvested grain and lentils in the fields.

The 'agricultural package' was nearly ready and was soon to conquer the whole of Europe. However, one important invention was missing. This arrived on the island about 7,500 years ago.

CHAPTER EIGHTEEN

The Farmers' Westward Voyages

All signs of human habitation on Cyprus ceased abruptly about 8,000 years ago. No one knows exactly what happened, but it looks as if everyone disappeared. Either they died, or they decided to leave the island. A possible explanation is the period of severe cold that struck north-western Europe 8,200 years ago. For over 100 years, the climate became a great deal colder, and over a 30-year period the average temperature sank by several degrees.

The cold was accompanied by drought, which must have been catastrophic for Cyprus. The landscape remains very dry today, leaving hardly any margin for farmers if it were to become any drier.

After its inhabitants disappeared, the island seems to have remained uninhabited for several centuries. However, about 7,500 years ago a new wave of settlers arrived. They came from the mainland by boat, like their predecessors, with a cargo of seed and domestic animals. But now there was a considerable and significant difference. The new generation of Cypriots mastered the art of making ceramic vessels. These are known as Sotira ceramics after the place where the first potsherds were found.

I visit the little village of Sotira one day in March. The landscape is still green after the winter's rain, and flowers bloom in abundance on the hillsides. A modern farmer is harvesting grain by tractor. In the courtyard outside the village café, four elderly men are having a chat. No sooner do they see me than they invite me over for a cup of Cypriot coffee, which is strong, brewed with the coffee grounds in the pot, and served on a tray with a glass of cool water.

Half an hour or so later, Carole McCartney joins me. She is going to guide me to Sotira Teppes, one of the first settlements on Cyprus where pottery was used. We climb through long grass up a steep slope, heading for the top. Both of us are out of breath, and I steal an anxious glance at Carole on hearing her strained breathing. She is of mature years and slightly on the plump side, and the last thing I want is for her to have a heart attack on my account. But she continues to stride tirelessly towards the hilltop.

The view from the summit is even more impressive than that from the older settlements further down which I visited earlier, the ones without pottery. You can look out over a wide area, all the way down to the sea a few kilometres away. A fine view seems to have been a high priority for Stone Age people. And another advantage of Sotira Teppes's location was safety. If attackers were to approach from the sea, the villagers would spot them in good time.

The remains of the little village are visible as stone walls in the grass. The first thing that strikes me is that the buildings are no longer circular, but square. The corners of the stone walls, however, are slightly rounded, not set at right angles as in our modern buildings.

Another difference is that these people no longer buried their dead under the floors of their homes, Carole McCartney tells me. Instead, they had special burial sites set slightly apart from their dwellings. They did, however, retain the custom of weighing the dead down with a large stone on their chests.

The ceramic ware produced on Cyprus during this period was coarse and porous. Vessels were painted red or white and patterns were applied with the help of combs. On a subsequent visit to Stockholm's Mediterranean Museum, I discover two small shards of the same type of pottery, displayed in a cabinet along with quantities of ceramics from later periods. There is nothing to explain that this is the first pottery to accompany the spread of agriculture westwards across Europe.

It is absolutely clear that the art of pottery came to Cyprus from the Middle East. But the question is: why did people on the mainland begin to make and fire clay vessels? As I mentioned earlier, the world's oldest ceramic vessels have been found in Japan, China and eastern Siberia. They were used by hunters and fishing folk at least 20,000 years ago. Even earlier than that – over 30,000 years ago – hunting people in what is now the Czech Republic made figurines of fired clay. So the first pottery had nothing to do with agriculture. Two of humankind's decisive technological advances were wholly independent of one another to begin with.

Somehow, farming people in the Middle East about 9,000 years ago realised that vessels made of fired clay were preferable to the stone vessels previously used. One theory is that this knowledge spread from eastern Asia, where it had existed for many millennia, to the Middle East. Another possibility is that the art of producing ceramic vessels reached the Middle East from Africa. There are finds of clay vessels from Mali that date back about 11,500 years, while ceramics older than any known Middle Eastern pottery have also been found in southern Egypt.

The third possibility is that the farmers of the Middle East invented ceramics themselves, independently of both East Asia and Africa. This was not necessarily such a huge step forward, as there are terracotta figures several thousand years older in existence. Early finds of this type also occur on Cyprus.

In the Cyprus Archaeological Museum in Nicosia, I look at a series of displays showing how vessels of various kinds developed over the millennia. The earliest stone vessels definitely look heavy and cumbersome. Ceramic vessels must have been much more practical, and, above all, considerably easier to produce once people had a basic grasp of the technique.

It is cool and comfortable inside the museum, which provides an excellent account both of Cypriot history and of how the world's first agriculture developed. You can follow the descriptions of how wild pigs gradually developed into the domesticated variety; how goats were initially left to run wild in the mountains, but later domesticated to provide milk in the villages; how people tried using sheep as a source of milk, but later reared them more for their meat and wool; and how difficult it seems to be to raise generation after generation of domestic cattle, especially in a dry environment.

Sometimes domestic animals died as a result of poor animal husbandry and inbreeding. In the Bible, it is by no means coincidental that God instructs Noah to take seven males and seven females of every type of animal with him in the ark. God was simply giving Noah a basic tip on how to build up a genetically viable pool of animals. It must have taken thousands of years to acquire such knowledge.

Sometimes people died from privations such as disease, drought or famine. At least once, and possibly several times, the entire population of Cyprus disappeared. Such was life during the first two millennia of farming: there were some harsh reversals of fate, and some successes. I could use the expression 'two steps forward and one step back' – but what do we mean by 'forward' and 'back'?

It is important to understand that people never had any long-term plan of 'inventing agriculture'. Rather, they solved problems that occurred in everyday life. They tried things out and found methods that worked. Agriculture – the most revolutionary development in the whole of human history – just happened along the way.

***

On our way back to the valley, we pass by the little café in Sotira again. The chairs are empty now. The old men who invited me for coffee seem to have gone home for lunch. I reflect that they, and other people living on Cyprus today, are descended in part from the very settlers who came here by boat several thousand years ago. They may even be related to the people who lived in the little hilltop village – Sotira's first inhabitants, who came by boat, produced ceramics patterned in red, built square houses with slightly rounded corners and buried their dead with large stones on their chests.

There is much evidence to suggest that the first farmers who produced pottery actually survived and stayed on Cyprus, and that an unbroken line links the Sotira culture to today's Cypriot people. However, the Stone Age settlers have, of course, intermingled with others over the millennia.

New waves of people have come to Cyprus one after the other: the first coppersmiths, the ancient Greeks, the Romans, Byzantines, Crusaders, Arabs, Persians, Ottomans, British, Turks, and so on.

Each era and wave of immigration has left its mark. This is manifest in the island's culture, which can be seen both in the facades of houses and in Cyprus's many museums. And today researchers can even detect the patterns revealed by the DNA of the current Cypriot population. Nearly a quarter of the people living on Cyprus have mitochondria from the H haplogroup – one of the groups typical of early farmers, and the group to which my paternal grandmother Hilda belonged.

What is more important, and scientifically more significant, are the recent more detailed analyses of nuclear DNA. These studies confirm that even today the population of Cyprus has an unusually large share of DNA from Europe's first farmers. However, the DNA signal from Europe's first wave of farmers is even clearer in a number of other places around the Mediterranean.

For nearly two millennia, Cyprus was the westernmost outpost of the first farmers. However, some 9,000 years ago seafaring farmers began to travel westward across the Mediterranean, in boats loaded with seed and domestic animals. Traces of farming from that time can be found along the coast of Turkey, on Crete and other islands belonging to modern-day Greece, and in mainland Greece.

A few thousand years later, the farmers had reached the east coast of Italy, after which they continued to travel westward along the Mediterranean seaboard at an ever faster rate. By 7,400 years ago, they had reached the Atlantic coast of Spain and Portugal.

Researchers have made some calculations and attempted to reconstruct these boats and their voyages. They conclude that a group of about 40 people would have needed 5 to 10 male sheep, goats, cattle and pigs and the same number of females (about seven of each, that is, just as in the biblical story of Noah's ark). A new settlement would also have needed about

250 kilos (550 pounds) of seed. About 10 boats would have been needed for transport, or, alternatively, fewer boats making return trips.

These voyages are likely to have taken place in late summer, allowing the settlers time for sowing before the winter rains set in. The first half year must have been critical. Probably it was only milk from cows, sheep or goats that enabled them to survive at the beginning; otherwise, it is hard to explain how they coped initially.

The DNA evidence of these early voyages is clearly observable on the Italian island of Sardinia, the main reason for this presumably being its mountainous relief and relative inaccessibility. The first wave of farmers could labour and toil here relatively undisturbed for millennia. For cultural reasons, many of them married within their own group. As a result, the genetic material of the first colonists has not mixed with that of other groups to the same extent as elsewhere.

***

The Basques, too, lived in relative isolation in the mountainous areas on the borders of Spain and France. They have even managed to keep their mysterious language, which has long baffled researchers.

Basque is in a language group of its very own. It is not one of the Indo-European languages, such as Spanish, French, Swedish and most other European languages. Nor is it Finno-Ugric, like Finnish, Sami, Estonian and Hungarian. And it is not a Turkic or Semitic language either. It may conceivably be related to some extent to certain isolated languages in the northern Caucasus, and to the language once spoken on Sardinia. This, at any rate, is what some linguists have claimed, although they have met with a good deal of opposition.

Many older linguists suggested that Basque was the last remnant of an early European language spoken by hunters before the farming population started to expand. This theory appeared to accord well with the very first attempts to identify the history of populations on the basis of genetic markers. The Basques have a particularly high proportion of rhesus-negative blood cells. In the 1980s, this fact caused geneticists to suspect that the Basques were a remnant of Europe's original population of hunters.

However, whether an individual has rhesus-negative or rhesus-positive blood cells depends on a single gene. Today, researchers examine hundreds of thousands of genetic markers when comparing nuclear DNA. For instance, they can see that a large proportion of Basque adults have the capacity to break down lactose, which means they can drink fresh milk. This is a hereditary trait that indicates long coexistence with dairy cattle, sheep or goats. It is clearly not a trait typical of an ancient hunting people.

Moreover, linguists studying Basque have not identified any particular words that might suggest an ancient language spoken by hunters. What they have found is a wealth of old Basque words for phenomena typically associated with farming, such as cow, bull, sheep, goat, herd, milk, butter, milking, wheat, rye, threshing, milling, pitchfork, bean, pea, and so on.

The most detailed analyses today show that Basques are genetically similar to contemporary Spanish and French people. However, the DNA of the Basque population as a whole is actually slightly closer to that of early farmers excavated by archaeologists. In particular, the Basques show a resemblance to 5,000-year-old farmers from an archaeological site at Portalón in northern Spain. These farmers, in their turn, were a mix of an earlier hunting people living on the Iberian Peninsula and farmers who had come from the Middle East.

Modern Basques would appear to be the direct descendants of the farmers of Portalón. This implies that they have many of the traits of the original hunting population, just as geneticists believed in the 1980s, but also an unusually high proportion of the traits of early farmers. By contrast, they exhibit a lower proportion of the traits characteristic of subsequent waves of immigration, which have had more impact on the rest of France and Spain.

The previously controversial theory that the Basque language was a remnant of the languages spoken by early farmers has thus received fresh impetus. Linguists have good reason to examine the suggested similarities between Basque and languages from the northern Caucasus. The Basque Country and the Caucasus are mountainous regions on opposite sides of Europe. The ancient languages spoken by farmers may have been preserved in these areas, although subsequent waves of new languages have taken over almost everywhere else.

***

The seafaring farmers who spread out over the Mediterranean produced a special kind of ceramic vessel. Its distinctive feature is that the potters pressed cockleshells into the clay to make an attractive pattern. Cockles were formerly classified as *Cardium edulis*, and this culture is therefore sometimes referred to as the Cardium culture.

Some farmers, however, made pottery of different kinds. Nor did all the farmers take to their boats. Some of them made their way westward on foot.

CHAPTER NINETEEN

The Homes Built on the Graves of the Dead

Agriculture spread rapidly eastward and southward from its heartlands on the upper reaches of the Euphrates and the Tigris. As we have seen, the Mediterranean island of Cyprus was one of the first stops when the farmers began their travels westward. But it took over a millennium for farming to reach western Turkey as well.

A number of tentative attempts can be discerned dating back about 9,700 years. They were made by early farmers who had not yet begun to use ceramic ware. However, the first sizeable wave of westward migration got under way more than a millennium later, when pottery was well established. The most massive wave of migration started about 8,200 years ago.

This point in time coincided with the climate event that took place 8,200 years ago – the great pan–European period of cold caused by the melting of an iceberg in Canada, as a result of which icy meltwater ran out into the Atlantic, altering the course of the Atlantic currents. It was at this point that people disappeared from Cyprus for several centuries, probably because of drought.

In the past, archaeologists disagreed on whether climate change in the Middle East really affected the lives of early farmers. The latest evidence, however, strongly suggests that it did. German and Israeli geologists and others have drilled in the bed of the Dead Sea. The core samples they obtained have enabled them to study variations in water level, pollen and clay particles, and thereby to establish how the climate has changed, century by century.

A dramatic change can be seen that occurred 8,200 years ago. The composition of pollen altered. The proportion of typical Mediterranean vegetation fell, while plants typical of deserts and dry steppes became far more common. At the same time, there was a dramatic drop in the level of the Dead Sea. The only possible explanation is a drastic fall in precipitation. Large areas that had once been suitable for cultivation dried out, becoming deserts. The Turkish archaeologist Mehmet Özdoğan believes the change in climate drove people away.

Today there are over 30 archaeological sites from the era when farming arrived in western Turkey. The finds from these excavations confirm that some farmers went to sea and followed the coasts westwards, while others made their way on foot overland. To simplify matters slightly, there were two distinct groups with slightly different cultures.

The seafarers generally constructed round buildings. They decorated their pottery by pressing seashells into the clay, and they buried their dead in separate burial places on the outskirts of their villages.

Those who made their way through inland Anatolia, on the other hand, built square houses. They used monochrome pottery that was generally undecorated and buried their dead under the floor in their homes.

**

Çatalhöyük, a flagship among archaeological sites in inland Anatolia, has been declared a World Heritage Site by UNESCO. It is sometimes called 'the world's first town', as there were several thousand people living there a full 9,000 years ago. The dwellings were square and built very close to one another – so close that there was no room for streets that would have enabled the inhabitants to move around. To reach their homes, they had to walk over their neighbours' roofs.

The only way of entering their dwellings was to climb down through an opening in the roof. The same aperture also served as a chimney, the oven being placed directly underneath the opening to allow the smoke to escape. There were probably ovens on the roofs as well, where Çatalhöyük's inhabitants seem to have spent much of their time during the warm part of the year.

Calling Çatalhöyük a 'town' is somewhat misleading. It was more like an unusually large village; although archaeologists have searched for many years, they have not found any communal structures. No shared storehouses, temples or burial places were built there, for example. Instead, each individual dwelling fulfils all these functions.

Each family had a small cupboard in which grain and lentils were stored, in woven baskets and grain containers of mortar and clay that were attached to the wall. Their main room contained a structure reminiscent of a shrine, which was kept particularly clean and well cared for. Some families adorned their shrine by incorporating the skulls of wild oxen directly into the wall. The horns of the oxen were painted with red ochre so that they stood out in all their splendour against the whitened walls. The red pigment was also used for mural paintings depicting creatures such as vultures, aurochs and leopards.

Often, the main room also contained the skulls of dead family members that were left lying out in the open. Their skeletons, on the other hand, were buried under the floor. The families living in Çatalhöyük built little platforms over the graves. We cannot rule out the possibility that they slept on these platforms – a few inches above their ancestors.

The walls, platforms and floor were daubed with different grades of clay and then limewashed, presumably several times a year. The white limewash made the rooms reasonably light, even though the sun's rays only filtered down into the interior through the little skylight.

The rooms would have had to be limewashed frequently because of the smoke and soot that billowed out of the ovens. And even though people kept their own dwellings whitewashed, clean and neat, they dumped rubbish, leftovers and human waste in the empty spaces between the houses, where the stench must have been abominable.

About once every hundred years, families built new houses on top of the old ones. The new house was nearly always identical to the old, with all its walls, both external and internal, in exactly the same positions. This could continue for 1,000 years – more than 30 generations.

Society thus seems to have been very conservative, with the ancestor cult playing a central role. However, life in Çatalhöyük was also very egalitarian as regards both social status and relations between the sexes. The homes of different families do not reflect any obvious difference in wealth. Analyses of skeletons and graves show that men and women ate food of the same quality and had equal status.

There have even been theories suggesting that Çatalhöyük may have been a matriarchy. Some researchers in the past believed its people mainly worshipped a goddess, given the numerous finds of terracotta figurines and drawings of plump female figures. The rotund terracotta figurines are very like those made by groups of European hunters all the way back to the Venus of Hohle Fels, nearly 40,000 years ago.

Such theories have encouraged contemporary followers of goddess cults to make pilgrimages to the site. Not infrequently, there have been cultural clashes between goddess–worshippers from elsewhere and people from farming villages near the excavations, who often have traditional Muslim values.

Ian Hodder, the archaeologist who has directed excavations in recent years, does not share the view that Çatalhöyük was governed by women who worshipped special fertility goddesses. Women could certainly play key roles, but so could men. Sexuality and fertility were important, but that applied to both sexes, according to Hodder.

The staple foods in Çatalhöyük were cultivated grain and lentils. There seems to have been a veritable plague of mice, which often nibbled at the grain containers in the home. Yet there are no signs of any cats. On reflection, I can well imagine that cats would have stayed at home in the regions where agriculture first emerged. Cattle, sheep and goats can be driven long distances, while dogs follow of their own accord. But cats do as they please.

Sometimes the villagers held great feasts at which they would grill meat from wild animals. There are a number of murals in which bearded men with quivers and leopard-skin loincloths are attacking aurochs and wild boar. Both the aurochs and the wild boar are males and are depicted in a state of sexual arousal. The terracotta art objects that have been discovered – apart from the female figures – nearly always represent wild beasts such as leopards and wild oxen. Sheep and goats are hardly ever depicted, although finds in the dwellings show that people ate such animals far more often than wild oxen.

Wild animals seem to have played a more important role than domestic ones in the imaginative world of Çatalhöyük's inhabitants. Although these people had lived mainly from farming for a few millennia, the ancient mythology of the earlier hunting culture lived on.

Çatalhöyük shows how two millennia of agriculture altered many aspects of life. Settlements could be much larger, accommodating 10 times as many people as in the past. People's diet contained more grain and therefore more carbohydrates, which sometimes led to caries. Life seems to have revolved more around the individual's own family and immediate forebears, rather than around the group as a whole. However, many aspects of the old hunting society remained. There were the same mythological wild animals and similar ritual female figures (whether or not we call them goddesses).

The people of Çatalhöyük seem to have lived in isolation to a large extent, and they were essentially self-sufficient. But they did import certain products from distant places. One of these was obsidian, a type of volcanic rock brought from mountainous areas 170 kilometres (110 miles) away. However, there is nothing to suggest that the newly arrived farmers in inland Anatolia met any local hunters – either in Çatalhöyük or anywhere else. In coming to Anatolia, they appear to have colonised an uninhabited land.

The migrants who made their way to the Sea of Marmara, the area that is now Istanbul and the shores of the Black Sea, had different experiences. These regions of course, had been inhabited since earlier times. The newly arrived farmers and the older hunting population must have been brought together here.

To judge from the remains Turkish archaeologists found, this went very well. Buildings, graves, tools and diet all indicate that groups of farmers and hunters merged, apparently in a peaceful fashion. This can be seen, for example, from the archaeological sites at Yenikapı and Pendik, near Istanbul.

Apparently there were no local hunters on the shores of the Aegean Sea, in the area where the port city of İzmir is now situated. But another kind of cultural encounter took place here. This was the meeting place of farmers who had migrated on foot across central Anatolia and farmers who had navigated along the coastline. In other words, migrant farmers from the hinterland and coastal farmers who had arrived by boat began to live together.

Sites such as Ege Gübre and Yeşilova near İzmir show how these two farming cultures merged. This fusion may quite conceivably have played a decisive role in the history of Europe. Presumably the coastal areas around İzmir were the starting point for the 'island-hoppers who brought civilisation to Europe', as some newspaper headlines expressed it.

'Island hopping' is a kind of holiday that involves travelling from one Greek island to another by ferry. I know people who spend their holidays this way. It now turns out that island-hopping farmers got about in a similar way all of 8,000 years ago. The newspaper headlines referred to a Greek study including DNA analyses of people living in Turkey, Greece and other parts of Europe today. The Greek researchers can demonstrate a clear pattern: people in central Anatolia are noticeably closely related to people living on Crete and the Greek islands known as the Dodecanese (which lie in the Aegean Sea between İzmir and Crete). These places also have a genetic link with mainland Greece and Macedonia to the north.

A number of other researchers would like to see more sound evidence in the form of DNA from archaeological finds. However, the Greek researchers themselves have already drawn their conclusions: one of the main routes along which farming came to Europe ran through inland Anatolia, then via the Greek islands to mainland Greece and northwards over the Balkans.

So it seems my foremothers on my paternal grandmother Hilda's side began their life in Europe as island-hoppers in the Aegean Sea.

***

The pioneers of European agriculture would have taken a range of different routes, including the one along the shores of the Mediterranean and the Black Sea. But I shall follow the route running northwards from the Aegean Sea and mainland Greece to the region around Thessaloniki, up along the River Struma and the River Vardar, through Macedonia, Serbia, Bosnia and Albania, and then northwards towards Hungary. This seems to have been the most important route by which agriculture reached Europe.

It is also the route taken by my own foremothers, Grand-mother Hilda's grandmother's grandmother's grandmother's grandmother's mother in the direct line, going back about 300 generations. The pattern on the map the family history research firm gave me is strikingly clear.

My paternal grandmother Hilda belonged to the subgroup of haplogroup H known as H1g1. On checking the genealogy researchers' register, I see that the most south-easterly instances of this haplogroup are to be found on the Peloponnese Peninsula in southern Greece and in the area around Athens. Genealogy researchers with the same DNA variants also crop up in Macedonia, Bosnia and Serbia, along the routes taken by Europe's first farmers on their way northwards towards central Europe.

When I began to plan this book, I asked the researcher Wolfgang Haak to comment on the degree of dispersal of haplogroup H1g1. He immediately observed that it seemed to be a typical marker of the first wave of migration of farmers into central Europe. Several subsequent DNA results confirm that theory. Today, conventional archaeology, researchers' DNA analyses and the private DNA analyses that I and other people researching their genealogy have commissioned all point in the same direction: a group of early farmers 'island-hopped' from the west coast of Turkey to mainland Greece before beginning the trek northwards towards Hungary.

Their arrival may well have been the most influential event ever to take place in this part of the world. Life in Europe would never be the same. After 30,000 years of the hunters' egalitarian, nomadic lifestyle, agriculture began to push Europe towards more private property, permanent year-round dwellings and greater class differences – in short, towards what many people call civilisation. It took a few millennia, but the first farmers to land in Greece took the first steps.

Having arrived in western Hungary, the first farmers developed the Starčevo culture. This was of decisive importance for the subsequent history of central, northern and north-western Europe. The Starčevo culture was the forerunner of the Linear Ware culture, which would later disseminate agriculture and dominate much of Europe, from France to northern Poland, for over a millennium.

There are detailed DNA analyses from Hungary that cast light on the first farmers in Europe. I hear about the results at a major annual conference organised by the European Archaeology Association. The conference is held in the Czech city of Pilsen, which is best known for its classic beer. That seems fitting to those of us who believe that agriculture and civilisation were driven largely by feasting and plentiful beer.

CHAPTER TWENTY

Clashes in Pilsen and Mainz

Pilsen is unseasonably hot for September. I am far too warmly dressed. My nylons are damp, the heat makes my feet swell and I discreetly kick my shoes off under the table, to get through the afternoon. I have taken a seat right in the middle of the front row so as not to miss a word of the very latest DNA findings about Europe's earliest farmers.

We are in the city's university, in one of the smaller lecture theatres. It is already full to bursting when the seminar begins. Participants fetch chairs from other lecture rooms. There are people sitting on the floor and squeezing in through the open doors. The organisers have seriously underestimated the degree of interest in DNA technology as a new tool.

The sun beats down on the windows, and the heat becomes increasingly oppressive – but it is nothing compared to the heated emotions of the participants. DNA in archaeology can clearly make feelings run high, both in Sweden and in central Europe.

'Even if these are the results we get from DNA research, that still doesn't make us migrationists!' cries an overwrought archaeologist. This is the first time I have heard the word 'migrationist', which is clearly a term of abuse. It refers to the theory that agriculture was disseminated largely by migrating farmers – precisely the conclusion currently being reached by one DNA study after another. These results are clearly viewed as being not quite respectable.

The word will crop up time and again during the event. It gradually dawns on me that for many archaeologists of the older generation, 'migrationism' has disagreeable ideological connotations and associations with Nazi ideas of a master race. But I see absolutely no sign of any such ideologies among the Hungarian researchers presenting their results. On the contrary, the research team leader, a professor of archaeology from Budapest called Eszter Bánffy, goes out of her way to give a warning about the way in which certain political forces in her home country misuse genetics. She is referring to racist propaganda against Roma and Jews, which sometimes deploys pseudoscientific arguments drawn from genetics.

As far as I can judge, the work done by Eszter Bánffy's doctoral students in cooperation with German scientists is nothing short of groundbreaking. A pale young man called János Jakucs explains how he has collected DNA samples from over 700 skeletons representing the first millennium of agriculture in Hungary and the neighbouring countries of Serbia and Croatia. He has visited museums and universities, negotiated and coordinated, and organised radiocarbon dating. Several other doctoral students report on their part in the overall project.

The main piece of news is that they have extracted mitochondrial DNA from the first farmers to arrive in Hungary, starting some 7,800 years ago. They have managed to analyse most of the 700 skeletons collected by Jakucs.

As I explained earlier in the book, virtually all analysed individuals from Europe's oldest hunting population – from Spain to Russia – belong to one variant or other of haplogroup U. Group U5, which I inherited from my mother and my maternal grandmother, is a typical hunter's lineage.

However, quite different variants are dominant among the farmers who appeared in Hungary 7,800 years ago: haplogroups such as N1a, T2, K, J, HV, W and X. They also include H, now the commonest group in western Europe, to which my paternal grandmother's maternal lineage belongs. As I have mentioned, both H and K have also been found in DNA analyses of the earliest farmers from Syria.

The most important results come from an archaeological site called Alsónyek-Bátaszék, which lies on the banks of the Danube in south-west Hungary. This was excavated between 2006 and 2009 in connection with the building of a motorway. In Alsónyek-Bátaszék, Hungarian archaeologists have found the most extensive vestiges ever of Hungary's first farming culture, the Starčevo culture.

The hours pass, the sun beats down on the windowpanes, and the room temperature rises – both the one you can measure with a thermometer, and the heat generated among the seminar participants. Many archaeologists who rely on conventional methods are extremely sceptical of what they view as 'migrationist' ideas – and, indeed, of DNA technology *per se* as an archaeological tool. At the same time, a conflict between different camps of DNA researchers is also emerging with growing clarity.

The Hungarian archaeologists and their German partners use basic DNA technology. Their testing is mainly restricted to mitochondria, which account for only a very limited proportion of human DNA, and they conduct only low-resolution analyses. Even the tests on my mitochondria and those of my paternal uncle, which I commissioned from a firm specialising in tests for family history researchers, are higher resolution.

There are also a few DNA researchers in the Pilsen lecture theatre who use a more sophisticated technology. By analysing nuclear DNA, they are able to obtain a more complete picture of the genetic history of individuals and populations. The researchers who analyse nuclear DNA believe that the picture given by those who work on mitochondrial DNA is oversimplified, too black and white. They are keen to emphasise the nuances that only their technology can detect. One German researcher working on nuclear DNA is particularly contemptuous of those whose purview is mitochondrial DNA. I get talking to him on the tram back into the city centre once the seminar is over for the day.

'Guido Brandt is an idiot,' he says. I am rather taken aback at his plain speaking, given that we have only been acquainted for a few minutes. My impression of Guido Brandt, a German doctoral student working with the Hungarians, is that he is anything but an idiot. It is just that he uses a comparatively simple DNA technology. The strength of Brandt's findings, and those of the Hungarians, is not the technical sophistication of their DNA technology. It is the large number of samples from locations and periods that are critical for the dissemination of agriculture in Europe.

***

A few weeks later I gain a deeper understanding of the conflict. I visit Guido Brandt in his home town, the German city of Mainz. I interview him in the chilly hut where he and his fellow doctoral student have been working since the university closed their institution.

Only a few weeks previously, Brandt was the first author of a study published in *Science*. Yet the university chose not to issue a press release. Instead, the press release came from a university in Australia, where a number of co-authors are based, and the National Geographic Society, which had provided financial support. The information department at Mainz's Johannes Gutenberg University even denies there is anyone called Guido Brandt at the university, although I have enquired after him on several occasions.

It transpires that he and a fellow doctoral student found themselves caught between two feuding professors. The two young postgraduates made a tactical error in betting on the wrong horse.

The university's press department also denies that the Johannes Gutenberg University now has anything to do with Kurt Alt, the former professor of anthropology in question. But he, too, exists nonetheless, and I meet him in the rooms where he and his secretary are in the process of winding up his long career. They are busy trying to empty a set of bookshelves that go all the way up to the ceiling and packing tons of books into boxes for removal.

The interview is not particularly informative. Alt, who was a dentist before making a career in anthropology, is clearly not very knowledgeable about DNA technology and the latest archaeological findings. Yet he was a pioneer in the field, being quick to recognise the opportunities it presented and to employ people with more extensive knowledge of DNA.

Left: The head of a lion and the body of a man: the Lion Man was carved out of ivory from a mammoth's tusk. This and other figurines from Swabia are the oldest known examples of figurative art.

Below: This preserved flute from the cave at Hohle Fels is made from the bone from a vulture's wing. The person who made it incised markings to ensure that the holes were pierced in the right positions.

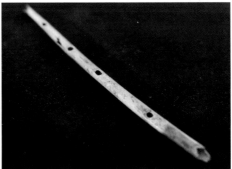

Below: The Neanderthals had already been living in Europe and Asia for several hundred millennia when modern people arrived and took over.

Above: The first farmers in Germany built this solar observatory at Goseck over 7,000 years ago. This means Goseck is considerably older than the better-known stone circle, Stonehenge. At the midwinter solstice, you can see the sun rise and set through two gaps in the wooden palisade.

Left: The people who used Göbekli Tepe dragged huge blocks of stone to the site. They carved images of scorpions, birds and other animals, and erected the stones in the round buildings where they worshipped.

Left: Mammoths are the most frequently occurring motif in the Rouffignac cave. Over 150 such pictures have been found.

Right: Ötzi weighed only about 50 kilos, but he was strong and wiry. The museum in Bolzano, Italy, displays the model with a bare torso. However, a striped cloak made of goat hide, a loincloth of goat leather, a bearskin cap and a raincoat made of grass stalks were among the finds.

Below: Brother, sister and betrothed? The three young people in Dolni Vestonice's triple grave are the oldest found to belong to haplogroups U5 and U8. All of them died about 31,000 years ago.

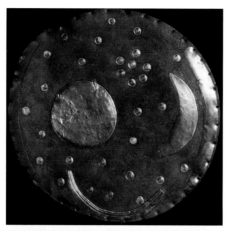

Left: The Nebra sky disc is made of bronze, with pieces of gold plate attached. It could be used by an expert to help predict the summer and winter solstices and synchronise the lunar and solar years.

Left: At the archaeological site of Khirokitia, archaeologists and craftsmen have reconstructed some of the round buildings where Cyprus's first farmers lived.

Below: Stonehenge seems to have been used mainly at the winter solstice. Farming people had been erecting similar circles of wooden posts almost since they first arrived in England. But the world-famous megaliths were added later on, some of them at the dawn of the Bronze Age.

RAPID CHANGES AFTER THE ICE AGE

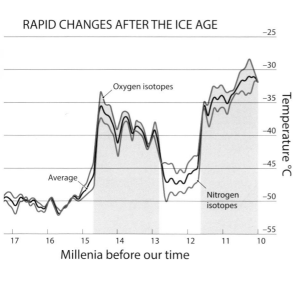

Oxygen isotopes

Average

Nitrogen isotopes

Temperature °C

Millenia before our time

Left: While the Ice Age was gradually slackening its grip on the globe, regional temperature changes could be dramatic. The curves illustrate analyses of cores from the Greenland ice sheet.

Below: DNA analyses shows the dispersal of people from the Yamnaya steppe culture, first towards the Corded Ware culture of central Europe and subsequently towards the Bell Beaker culture of western Europe.

THEY CAME FROM THE STEPPE

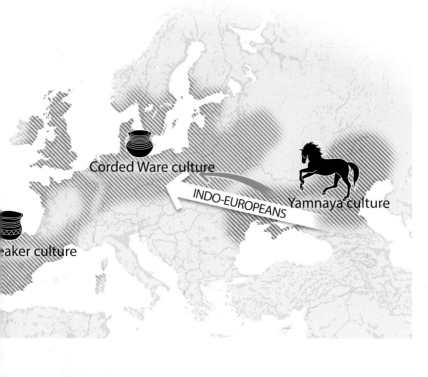

Corded Ware culture

INDO-EUROPEANS

Yamnaya culture

•aker culture

R1a CLANS

The Indo-European pastoralist culture of the steppes included men belonging to haplogroup R1a. During the Bronze Age, they had a very large number of descendants who spread throughout Europe and down to south Asia.

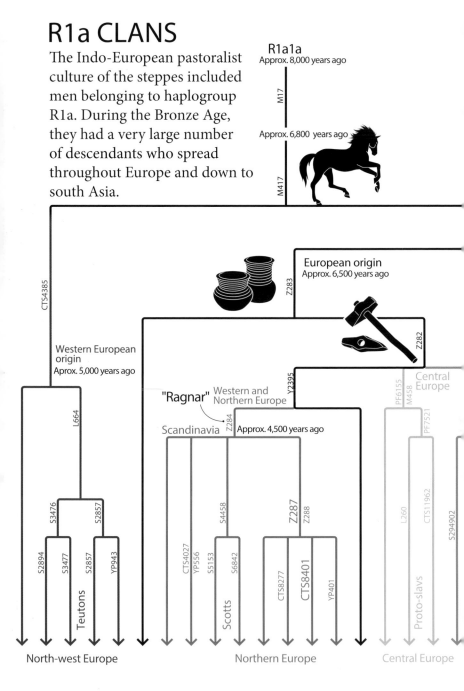

R1a1a
Approx. 8,000 years ago

M17

Approx. 6,800 years ago

M417

CTS4385

European origin
Approx. 6,500 years ago

Z283

Z282

Western European origin
Aprox. 5,000 years ago

L664

Y2395

Central Europe

PF6155

M458

PF7521

"Ragnar" Western and Northern Europe

Z284

Scandinavia

Approx. 4,500 years ago

S3476

S2857

S4458

Z287

L260

CTS11962

S2894

S3477

S2857

YP943

CTS4027

YP556

S5153

S6842

CTS8277

CTS8401

YP401

S294902

Teutons

Scotts

Z288

Proto-slavs

North-west Europe

Northern Europe

Central Europe

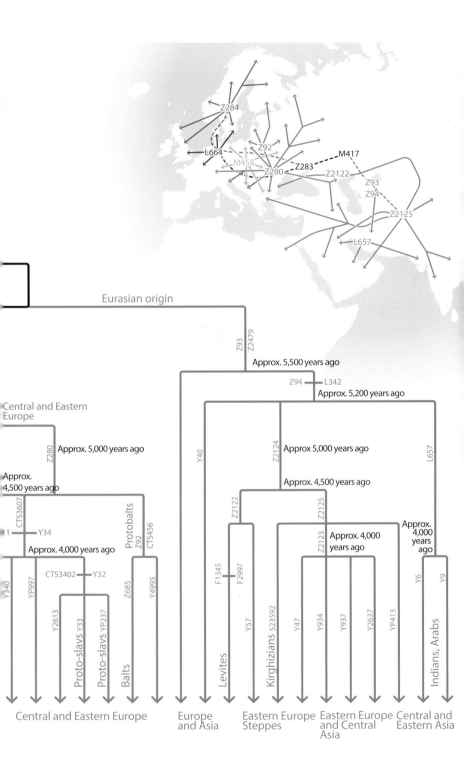

Right: There is a guesthouse today next to the site where the first Cro-Magnons were discovered. The house is built into the cliffs, just like the huts of the Cro-Magnon people. Here I am in front of the house.

Right: My mother, Anita, aged about twenty.

Below: My paternal grandmother, Hilda, with her mother, Elin, and her daughter, Gunnel. The little girl in the picture is my cousin Kristina. Today she is a grandmother herself, and her daughter's daughter has the same mitochondria as the women depicted here.

That was where it all went wrong. Alt and a newly recruited expert fell out in less than half a year. 'I've never seen a personal relationship break down so quickly,' says one of the researchers who saw it happen. That researcher has now left the whole toxic DNA field and is working in another field at a museum in Mainz. Another former member of the team has moved to Australia.

But Guido Brandt and his fellow doctoral student are still there, working away in a chilly hut and trying to bring their theses to a successful conclusion in spite of everything. And it all seems to be going very well. The study recently published in *Science*, of which Brandt is the first author, deals with 364 individuals from the Stone Age to the Bronze Age in the German region of Saxony-Anhalt. It covers more individuals whose DNA has been analysed than all the archaeological studies ever published taken together. Even if his work is restricted to low-resolution analyses of mitochondrial DNA, it is still quite an achievement.

The results confirm what earlier, less extensive studies have shown, as do the results from Hungary: all the early hunters belong to haplogroup U. They are classed as U5, like me, or U4 or U8. About 7,500 years ago, agriculture reached the region that is now Saxony-Anhalt. Archaeologists were already aware of this, partly because the linear pottery that gave its name to the Linear Pottery (LBK) culture made its appearance at exactly that time. Undoubtedly, local hunters would have joined them in the course of the journey. But it is only the lineages including haplogroups H and K that appear to have come all the way from the cradle of agriculture in Syria.

I also meet the 'winning horse' – the Johannes Gutenberg University's new professor of anthropology, Joachim Burger. His office is on the edge of Mainz's botanical gardens, in an attractive large building that the university recently had built for him. One of his young students shows me to his study and knocks respectfully on the door.

Joachim Burger's face stiffens when he hears I have also met Kurt Alt and Guido Brandt. He makes it clear to me that he has no intention of forgiving Brandt for what he regards as treacherous behaviour. In other respects he is friendly. He has made tea for us and lights a nightlight to keep the teapot warm. He gives me tips about wine cellars and restaurants in Mainz. The situation feels almost like a date, rather than an interview.

Naturally he wants to show me his newly built lab, which he describes as the most advanced DNA laboratory in the whole of Germany. But it is late afternoon, and the alarm system is so complex that we are unable to enter the lab. So Burger shows me a film about it instead, set to his own music; composing is his hobby.

Burger is a successful researcher by any standards. He collaborates with the main players in his field and has published a number of studies in leading journals. Some of his results, pertaining to the DNA of domestic animals and people, are dealt with elsewhere in this book. I will confine myself here to the study published in *Science* on the same day as Brandt and Alt's work – the one that the university's information department chose to focus on exclusively in its press release.

Burger says he is not really in favour of researchers analysing mitochondria in isolation, now that there are so many better methods. Yet the study he published in *Science* deals with mitochondria. More specifically, it covers mitochondria in combination with isotopes. It is very ingenious and interesting.

Comparing different isotopes of carbon, nitrogen and sulphur provides insight into people's diet. Burger and his team examined 29 individuals buried in a cave called Blätterhöhle. They date back to the time just before agriculture arrived in the region that is now Germany and a few millennia after that.

The oldest individuals are quite clearly hunters. The isotopes show they lived mainly on fish and meat from wild animals, and they all had mitochondria belonging to haplogroup U. Eight of the individuals from more recent times are quite clearly farmers. They belong to typical haplogroups such as H and J, and the isotopes indicate that they ate meat from domestic animals such as sheep and cattle.

Hitherto, the picture was blurred by 12 individuals who lived at the same time as the farmers and were buried in the same cave, but whose mitochondria were typical of hunters, belonging to groups such as U5. Now the isotopes show that they actually lived different lives from the farmers, despite being buried at the same spot. The hunters lived mainly on fish.

Burger's study thus indicates that groups of hunters and newly arrived farmers lived side by side for at least a few thousand years. They clearly had a good deal of contact with each other. Apart from the fact that they used the same cave as their burial ground, the farmers seem to have obtained all their stone tools from the hunters.

Over centuries and millennia, hunters and farmers intermingled and merged. In addition, new waves of immigrants arrived. Guido Brandt's major *Science* study of the mitochondria of people living in Saxony-Anhalt over four millennia shows clear signs of such changes. It points to subsequent waves of farming people who migrated into the region from the north and the east.

CHAPTER TWENTY-ONE

Sowing and Sunrise

I travel to the German city of Halle to learn more about the first farmers in Germany, those known as the people of the Linear Pottery (LBK) culture, after the patterns on their pottery.

The Museum of Prehistory in Halle, founded in 1819, is said to be the oldest of its kind in the world. A solid edifice with huge sandstone pillars, it embodies the weight of tradition. Though the building itself is old, the displays are state of the art, accessible and based on sound educational principles.

One of the display cases contains a small wooden model of a typical building from the time of the Linear Pottery culture. It is a wooden longhouse built on posts, its outer walls attractively decorated with red and white zigzag patterns. Pressing a button next to the display case brings forth sounds that evoke life in the longhouse: intermingling voices, children's laughter, millstones grinding, the lowing of cattle and the cries of a goat being slaughtered.

A similar scene is depicted in a large picture on the wall painted by a talented artist attached to the museum. It shows the farmers of the Linear Pottery culture clad in simple hide garments, with red patterns on both their clothes and their skin. Bearded men are slitting a goat's throat. Men and women alike have long hair. Dogs, cows, sheep and goats wander about freely, but the people can avoid their animals by climbing into their stilt-house. Its walls are adorned with the skulls of cattle and geometric patterns painted in red pigment.

Bernd Zich, an archaeologist and the director of one of the museum's departments, guides me through the exhibitions and tells me about the first farmers in the region we now know as Saxony-Anhalt.

They arrived 7,500 years ago. The traces they left behind them stand in strong contrast to those of the hunters already living in the area. From the very beginning the farmers were fully equipped with a complete 'package' of agriculture. They kept cattle, sheep and goats, and they grew wheat and leguminous crops.

Zich believes the conceptual world of the farmers of the Linear Pottery culture was very much based on astronomy. Growing crops made it especially important to keep track of the sun and the solar year. They needed to know the right time for sowing. As farmers, they needed to plan ahead far more than the hunters had done. It was not enough just to rely on the path of the moon, as the lunar year does not correspond exactly to the solar year. Relying solely on the path of the moon would put you a whole month out after three years. That was why the vernal equinox, the summer solstice, the autumn equinox and the winter solstice were hugely significant.

In the areas inhabited by the farmers of the Linear Pottery culture, archaeologists have found the remains of several large structures that they call solar observatories. The best known of these is Goseck. It has been reconstructed to show how it once looked. On my second visit to Halle, I hire a car and head out there.

It is late November, with about four weeks to go until the winter solstice. The weather is cold and windy. First of all I visit the castle, which houses a small museum. The lady selling tickets at the entrance looks almost shocked to have a visitor on a Sunday in late November. There are no other visitors to be seen inside – a pity, as there is much to be learned here about the world's oldest signs of astronomical observation. Goseck is considerably older than the Egyptian pyramids and its better-known English counterpart, Stonehenge. On a map, researchers have marked all the solar observatories that have been identified from the time of Europe's first farmers – a dozen of them. The one furthest south is in Hungary, while the most northerly one lies outside Berlin.

All the observatories resemble one another, though there are minor differences. The basic pattern comprises a series of circles – an embankment on the outside, followed by a deep ditch, with concentric rings of tall oak posts set close to each other inside. At Goseck, the outer bank seems to have been used as a burial place. Archaeologists have discovered several vestiges of human skeletons interred in the embankment. At other points in the circle, people lit fires and buried objects. Archaeologists believe the centre of the circle was used to make offerings. A drawing exhibited in the castle museum depicts people about to sacrifice an ox at sunrise, on the morning of the winter solstice.

After a dose of theory in the little castle museum, I drive a few kilometres to the actual structure, which lies in the middle of a large, windswept field. It is a few hundred metres' walk, and the winds are chilly. No doubt it would sometimes have been at least as cold at the time when Germany's first farmers used the observatory; after all, its raison d'être was to let them know when the winter solstice had arrived.

The circular embankment surrounding the structure at Goseck has a diameter of about 75 metres (245 feet). There is an opening at the northernmost point, where I enter – just as the first farmers must have done for their sacrificial ceremonies. Their dwellings were due north of the observatory, which appears to have served as both temple and burial place.

There are similar openings to the south-east and the south-west. This means I can stand in the centre of the circle and look straight out in three directions: north, south-east and south-west. Today you can see a house through one opening and a few trees through the other. But when the observatory was in use there must have been an unobstructed view of both sunrise and sunset.

At the time of my visit, the weather is not only cold and windy, but clouded-over into the bargain. Moreover, I'm here in the middle of the day, not at sunrise, so some of the charm of the spot is lost on me. The castle museum has several pictures by an expert photographer on display. That photographer – unlike me – timed their visit to ensure clear weather and the right time of day; the results are incomparably beautiful. The first rays of the rising sun, pointing towards a rosy horizon, gleam through the narrow gaps between the oaken posts. At that precise moment, you know the darkness of winter has begun to decline, and spring is on its way back. Once the moon has run its course a few times, it will be time to start sowing again.

Clearly, early hunting peoples also kept a close watch on the starry sky, not least in order to determine their position and direction. But it was probably the first farmers who started a cult of the sun and moon. People in today's Europe still experience aspects of that cult. We continue to celebrate the turning points of the solar year. For instance, our main feast day – Christmas, or Yule – is rooted in ancient festivities associated with the sun. The Bible is silent about the season of Christ's birth. It was early Christians who decided in the third century AD that the day of his birth should be celebrated at the midwinter solstice, on the same date as an ancient feast in honour of the sun.

Researchers have calculated the average speed at which farmers made their way towards Europe from the heartlands of agriculture on the banks of the Euphrates and the Tigris. If their speed had been unvarying, they would have moved about a kilometre (⅔ mile) a year. But that figure is misleading, as they are more likely to have migrated by leaps and bounds. Sometimes they moved very quickly, while at other times they remained in the same place for a long time.

The people of the Linear Pottery culture covered the distance from the banks of the Danube in Hungary to the area around Halle in Saxony-Anhalt, in central Germany, within just 300 years – about 10 generations. Throughout that time, and for a further thousand years, the Linear Pottery people settled in locations rich in loess. Loess is a soft, deep and fertile soil comprising fine particles blown by the wind during the last Ice Age.

At the time when the first farmers came to Europe, lush forests of hardwood broadleaved trees flourished in areas with loess soil. Lime trees, elms, ash trees and oaks had been shedding their nutrient-rich leaves for millennia. This foliage had rotted, forming layers of the finest mulch a metre (3¼ feet) thick. The dense foliage of the treetops kept out the light, so there was practically nothing growing on the ground apart from spring flowers such as hepaticas (blue anemones) and wood anemones. Basically, all the farmers needed to do was cut down the trees, press seeds into the earth around the stumps and leave them to grow. Experiments show that a skilled person can cut down a substantial tree in under an hour using a flint axe.

However, the boundary between loess and non-loess soils runs through northern Poland and northern Germany. Here moraine takes over, with thin sandy earth and clayey soils that are full of nutrients, but hard. The farmers stopped at this boundary for over 1,000 years. Different methods were needed to be able to grow crops on the soils further north.

Thin sandy soils lend themselves to swiddening (slash-and-burn cultivation), a technique based on felling the trees in winter, burning everything on the ground in spring, sowing and harvesting. Crops flourish in the ash, which is rich in nutrients, but only for a few years. Later on, weeds and undergrowth take over and livestock can graze on the cleared land, while the people move on into the forest and create new clearings.

Clayey soils can be excellent for growing crops for those with access to wooden scratch ploughs (ards). But ploughs do not plough by themselves; they need something to pull them. About 6,000 years ago oxen took on far greater significance. This can be seen from a series of finds of statues and images featuring them. Their sudden importance may be linked with the plough and the wheel.

Bernd Zich of the Museum of Prehistory in Halle found the world's oldest known wheel tracks as a young man working on excavations in Flintbek in the north German state of Schleswig-Holstein. The tracks are an estimated 5,500 years old. Archaeologists in Flintbek had already found the traces of a plough. Similar tracks – up to 5,700 years old – have now also been discovered in Poland, the Netherlands, Denmark and Sweden.

The oldest known image of a wheeled vehicle adorns a 5,600-year-old terracotta vessel in Poland. This was discovered in excavations at Bronocice and is now kept in Cracow's archaeological museum.

Zich believes that oxen and wheeled vehicles became so important that they rapidly acquired a mythological significance. He interprets some of the stone circles made by late Stone Age farmers as a kind of representation of such vehicles. When I first visited the Halle museum, the staff were in the process of setting up an exhibition about a stone circle representing a wagon.

Perhaps not everyone subscribes to the interpretation that the great Neolithic stone circles are actually a kind of wheeled vehicle. However, traces of wheels and ploughs undeniably began to appear about 5,700 years ago, and they radically altered the conditions for farming on different kinds of soil.

The same applies to the fine, soft loess soils. The first generation of people from the Linear Pottery culture in Saxony-Anhalt tilled fields the size of the relatively small garden of a detached house – about 500 square metres (5,400 square feet). One and a half millennia later, the fields had become tens of times larger. Oxen and wooden ploughs may partly account for this change.

***

With today's knowledge, it is impossible to be sure who invented the wheel and began to use ploughs with oxen as draught animals. Once the technology was there, it spread like wildfire everywhere between northern Europe and Mesopotamia. But in my view, northern Europe, particularly Jutland or northern Germany, might very well have been the birthplace of the wheel.

What came first – the chicken or the egg? Were farmers able to leave loess soils behind them and move northwards only once they had invented the wheel and started to use ploughs and draught animals? Or did they begin to use ploughs and draught animals out of necessity, when they had already migrated northwards to areas with harder soils?

The areas around the coast of the Baltic Sea and in southern Scandinavia had a number of drawbacks for a farmer. Not only was the climate harsher and the soil harder to till, but the coastline was already peopled by vigorous hunters. But one day, about 6,000 years ago, a number of farmers from the region that is now Germany took that step. And after that things moved very fast.

CHAPTER TWENTY-TWO

Farmers Arrive in Skåne

That they arrived by boat is beyond doubt. Setting off from northern Germany, they landed along the coastline of Skåne and Halland in southern Sweden, and, at the same time, on the Danish island of Fyn (Funen). We know Sweden's first farmers had boats, as they also reached the islands of Bornholm and Gotland at about the same time.

The oldest traces of agriculture in Sweden go back about 6,000 years. Within just a few generations, farming had spread all the way up to Bohuslän, the Oslo Fjord in Norway, and Uppland in eastern Sweden.

Sweden's first farmers loaded their boats with cattle, sheep, pigs, barley and several types of wheat. They used a specific kind of stone axe to fell trees, and they built longhouses of a characteristic type. They also used a particular kind of terracotta ware known as Funnel Beaker pottery (TRB), which developed out of the older Linear Ware (LBK) of central Europe.

The distinguishing feature of Funnel Beaker pottery, as the name suggests, is that vessels were often shaped like funnels, ending in brims that flared out to varying degrees. When laying their dead to rest, the Funnel Beaker people would often bury a beaker with this type of brim as grave goods. The vessel's shape meant it could be attached to a belt, like a hip flask.

It seems quite clear that Funnel Beaker vessels had a very special significance, which may have stemmed from their contents. Many archaeologists strongly suspect that the Funnel Beaker people used the barley they cultivated to brew beer, which would have been quaffed from these beakers. One clue is the numerous finds of burnt grain in their settlements. Roasting is a stage in malting barley for brewing.

The evidence would be even stronger if chemists had been able to detect traces of oxalate in Funnel Beaker vessels. This involves analyses like those conducted by archaeologists searching for traces of brewing at Göbekli Tepe in Turkey. If it were proven that Sweden's first farmers drank beer at their ritual gatherings, there would be a link running all the way from the place of worship at Göbekli Tepe, which dates back 12 millennia, to the Sweden of 6,000 years ago, and to celebrations in our own time.

Archaeologists from Skåne have put forward a nice idea: the newly arrived farmers may have asked the original hunting people round for a beer, as a conciliatory and fraternal gesture. Whether there is any truth in that we shall never know. We have no idea what feelings and relationships may have existed between the hunters of earlier times and the farmers who arrived subsequently. However, what research can show clearly today is that there were two quite separate groups to begin with.

For over 30 years, the predominant view held by Swedish archaeologists was that agriculture emerged in Sweden when local hunters decided to start tilling the soil and raising livestock. However, modern DNA research has shown that this narrative cannot be correct. DNA evidence shows clearly that agriculture arrived as a result of inward migration – with people who came by boat and on foot from distant Syria, via Cyprus, Turkey, the Balkans and central Europe, to northern Europe.

The Danish archaeologists Lasse Sørensen and Sabine Karg have demonstrated the same thing, though using quite different evidence. They have made a thorough and extensive compilation of excavations in Denmark and Sweden. Their study differs from previous ones in that it covers more sites – both on the coast and inland – and in the exacting standards they applied to radiocarbon dating.

As a result, a new pattern has emerged that is quite different from the one many archaeologists believed they could discern in the 1980s. The old view was based on a smaller number of excavated sites – nearly all of them on the coast – and less reliable dating, mostly based just on the layer from which buried objects had been retrieved.

Sørensen and Karg demolish most of the arguments that groups of hunters along the coasts of Skåne and Denmark engaged in farming early on. They show that bones claimed to have belonged to a cow may actually come from a wild ox. Grain and bones from domestic animals which, it was claimed, were over 6,000 years old and came from a group of hunters, are actually far more recent; it just so happens that they slid down to a lower level in the soil. Individual bones from domestic animals and impressions of grains in shards of pottery may have been ascribed to the correct layers. However, individual finds of this type may mean that hunters bartered goods for animals or grain from farmers further south, not that they began to farm themselves.

The overall picture that emerges if you consider all archaeological sites is that farmers began to arrive in Sweden six millennia ago, and that most of them settled in inland areas. The hunters continued to live along the coasts for many hundreds of years, during which they retained their way of life at least in part.

There would certainly have been a good deal of contact between these groups, and the boundary between them was not watertight. Some farmers also went hunting and fishing. Hunters were given domestic animals from time to time, or acquired them through barter. Sometimes farmers and hunters had children together. Little by little, the two groups merged completely.

***

Why, it is worth asking, did the farmers suddenly begin moving northwards into Sweden and Denmark?

One standard answer is that early agrarian cultures always had a need to expand. According to one hypothesis, farmers generally had more surviving children than groups of hunters. The farmers' young children had a major advantage: they could stay in one place. They were not obliged to move from one hunting ground to another, which meant long treks with unreliable access to food and clean water. Since more children survived to adulthood, farming settlements grew faster. However, their limited farming techniques meant they were unable to provide for the expanding population. Consequently, there were always people who had to move and find new ground.

There is nothing wrong with that hypothesis. However, it fails to explain why the farmers of the Linear Pottery culture came to a halt for 1,300 years at the northern limit of loess soils – and why their successors, the Funnel Beaker people, suddenly moved into northern Germany and Poland, and a little later into Denmark and Sweden.

Many archaeologists mention the flint in Skåne as a major attraction. Exceptionally high-quality flint can be found in Södra Sallerup, Kvarnby and Tullstorp, outside Malmö. Thousands of pits up to seven metres (23 feet) deep show that flint was already quarried on a large scale six millennia ago. Some of the first traces of farming in Sweden are very close to these flint quarries.

There are also traces here of flint axe production on a virtually industrial scale. The farmers would certainly have needed such axes to fell trees when clearing new farmland. And it seems more than likely that the high-quality flint in the Malmö area enticed the first farmers into Skåne. However, that does not explain why agriculture continued to spread northwards into Västergötland, reaching the Oslo Fjord and Uppland within just a few generations.

Another hypothesis is based on the fact that the sea level sank at about this time, as the climate in Europe grew drier. As a result, coastal meadows and river valleys emerged that had previously been under water. Grass and plants providing ideal grazing for the farmers' sheep and cattle flourished on this fertile land. Again, this may well be part of the story, but I remain unconvinced that the new meadowland provides an adequate explanation.

More persuasive, to my mind, are the ideas put forward by Bernd Zich from Halle, who links the spread of agriculture into northern Europe with the ox, the wheel and the wooden plough.

As far as I can see, there are two possibilities. One is that farmers started to use wooden ox-drawn ploughs over 6,000 years ago. They developed the yoke, and this technological change coincided with the upper limit of loess soils, somewhere in northern Poland, Germany or perhaps the Netherlands. Once the farmers had the necessary technology to cultivate heavier soil, they were able to continue their expansion northwards.

Alternatively, the expansion to the north may have come first, followed a few hundred years later by the new technology. When the farmers who had recently settled in northern Europe were obliged to work heavier soils, they thought of starting to use oxen as draught animals, and they developed the wheel and the wooden plough.

Whoever invented the wheel, it is clear that flint axes, oxen and cattle made their mark on the Swedish landscape. Their impact is still visible today. Dense forests of deciduous hardwood trees have given way to open landscapes – a consequence of 6,000 years of agriculture.

And at many locations in farming country you can still see formations of gigantic boulders that were dragged into position by oxen. These are particularly common on arable land in Västergötland, which I am to visit soon.

CHAPTER TWENTY-THREE

Ötzi the Iceman

B ut before going further and following the first farmers of Skåne on their journey northwards into Sweden, I make a detour southwards to the Alps. At a museum in the northern Italian city of Bolzano, I pay a visit to Ötzi.

'The Iceman from Tyrol', as he is also known, was discovered in 1991 by a German couple on holiday, the Simons. They came across the body while out hiking near the Ötz Valley, hence the Iceman's nickname. Assuming he was a mountaineer who had had a fatal accident, the Simons immediately rang the police and the Alpine rescue service. However, forensic experts quickly realised that Ötzi was very old – so ancient, in fact, that he was a case for archaeologists rather than forensics. He was, to be precise, 5,300 years old. For all those years he had lain frozen in the glacier until the ice was melted by unusually hot weather. As luck would have it, the German holidaymakers chose to turn aside from their waymarked path, so they happened to pass by just as Ötzi was beginning to thaw out.

And there was another coincidence: Ötzi had died just before it started to snow. Owing to the cold, the dry air and the protective covering of snow, which gradually turned into ice, his body was mummified by natural processes. Quite simply, he was freeze-dried.

Ötzi's exceptionally well-preserved body and the artefacts discovered around him make him one of the most informative finds ever in European archaeology.

His story is an eventful one. For example, it was long unclear which country had a rightful claim on him. Eventually, it was established that he had been found 90 metres (295 feet) from the border, in the Italian region of

South Tyrol – which is why he is now in the museum in Bolzano.

A few years on, researchers discovered he had been the victim of a murder, albeit one committed 5,300 years ago. Someone had shot him from the back with an arrow that struck him high up on his left shoulder blade. The arrow hit a blood vessel, injuring him so badly that he must have bled to death within a few minutes. He also seems to have suffered a blow to the head just before he died, possibly as his killer attempted to pull the arrow out.

All this can be learned at the museum in Bolzano, where the visitor can look through a glass pane at Ötzi himself, lying naked and dried out in a cold store, his skin brown and leathery. You can also see a full-size reconstruction that shows what he probably looked like when alive (please see central plate section). It is based on the appearance of the corpse at the time of discovery.

The model shows a man aged about 45 and about 1.6 metres (5¼ feet) tall. Though thin, at just 50 kilos (110 pounds), he is nonetheless strong and lithe. His eyes are brown. He has a dark beard, and his dark brown hair is long and wavy, falling at least to his shoulders.

The exhibition also displays garments found where the glacier had thawed. There is a striped cloak of goat hide and a loincloth of goat leather. Ötzi's leggings were also made of goat hide, with deerskin laces designed to keep the snow out. Ötzi's cap is made of bearskin, while his shoes had deerskin uppers and bearskin soles, and were lined with hay. His belt, made from the hide of a calf, contained several tools of flint and bone.

He had a flint dagger, a large bow and a quiver made of chamois hide reinforced with hazelwood. The quiver contained a number of half-finished arrows made of young shoots from guelder rose trees (*Viburnum opulus*). A birch-bark container seems to have been used to carry glowing coals. If the glow went out, he would have been able to make fire again by striking a piece of pyrite against a piece of flint and igniting a scrap of tinder fungus with the spark. He had on him all the tools for lighting a fire and keeping it going. There was also a woven net for trapping prey and a device for carrying the wildfowl he had killed. Researchers identified a mat made of grass fibres as a sort of raincape. They think a structure made of hazel twigs was a framework for a backpack, or possibly a pair of snowshoes.

The equipment bears striking similarities to that used by outdoor activities enthusiasts today. The materials used, of course, are quite different from today's glass fibre and synthetic fabrics that 'breathe'. There are no fewer than 18 types of wood in Ötzi's gear, plus hides, sinews and bones from various species of animal and fibres from several plant species.

His intestinal contents reveal that his last meal consisted of wheat, plant material and goat meat. His knee joints show signs of wear and tear, suggesting he led a vigorous life in the mountains. Nor are his teeth in perfect condition. The necks of his teeth are inflamed, for one thing. The chewing surfaces are worn, both by Ötzi's penchant for using his teeth as tools, and as a result of eating flour containing traces of millstone grit.

His lungs show traces of soot particles, showing he often sat next to smoky fires. He suffered from Lyme disease, spread by ticks, rheumatism, and an intestinal parasite called the human whipworm (*Trichuris trichiura*), which is borne by dirty water. He would often have suffered from pain in his joints and back.

This may explain why he had recourse to acupuncture. His skin bears tattoos shaped like tiny strokes. Many of them are at exactly the same points as those used by traditional acupuncturists. Presumably tattooing was a way of trying to dull the pain. Before Ötzi's tattoos were detected, acupuncture had been thought to be a purely Chinese tradition. However, there seem to have been similar ideas in Europe as far back as 5,300 years ago.

Earlier in his life Ötzi had broken some ribs and a nasal bone. His nails bear several horizontal furrows caused by illness, malnutrition or severe stress.

Isotopes in his teeth and bones indicate that he came from a valley about 60 kilometres (40 miles) to the east. However, he appears to have spent his adult life in a valley much closer to the spot where he was found.

There have been numerous speculations, some more well-founded than others, about Ötzi's identity. There is much to suggest that he was a herdsman who kept watch over his clan's livestock in the mountains. He may also have been the chieftain of a clan that was at loggerheads with other clans. Alternatively, he may simply have been a thief and bandit living the life of an outlaw – or maybe he was a shaman and coppersmith.

He cannot have been an ordinary, run-of-the-mill man, as his pack contained an axe with a copper blade. Ötzi died on the cusp of the Neolithic era – the farming Stone Age – and the period known as the Chalcolithic era (the Copper Age) in the Alpine region. Objects made of copper made their appearance at this time, but they were infrequent. A copper-bladed axe must have been a very rare and costly artefact.

Traces of arsenic in his body reveal that Ötzi himself worked as a coppersmith, which must have been a high-status craft. The ability to transform stones into gleaming metal with the aid of fire must have been regarded as something extraordinary – indeed, nothing short of magical.

Ötzi was the first prehistoric individual to undergo DNA analysis. In 1994, researchers were already able to analyse his mitochondria in general terms and establish his haplogroup. He turned out to belong to haplogroup K, which we now know to be a typical marker of the expansion of early farming in Europe.

It later emerged that haplogroup K was already present in very early Syrian farmers living 10,000 years ago. The same group was also found in farmers belonging to the Starčevo culture in Hungary 7,800 years ago, and it was common among the Linear Pottery farmers of Germany 7,500 years ago. Today, about 6 per cent of the European population belongs to haplogroup K. It is more common among Kurds, Druze people and Ashkenazi Jews. However, more detailed subsequent analyses have shown that Ötzi belonged to a subgroup of haplogroup K that has no other known members today. It is known as K1Ö, with Ö standing for 'Ötzi'.

In 2011, researchers also succeeded in analysing Ötzi's Y chromosomes, enabling them to unravel his paternal lineage. This revealed that his closest relatives in Europe's present population are to be found mainly on Corsica and Sardinia. Ötzi's Y chromosomes belong to haplogroup G, again a typical marker for the expansion of early agriculture in Europe. To be precise, he is part of a subgroup called G2a4, which occurs in less than 1 per cent of the population of Europe as a whole. However, in southern Corsica and northern Sardinia, nearly a third of all men belong to this rare haplogroup.

This study was followed in 2012 by a very detailed analysis of the Iceman's nuclear DNA, which enabled the researchers to make statements about a number of his inherited traits. They could confirm that he had brown eyes and wavy brown hair. Moreover, he was genetically disposed to furred arteries, as researchers had already observed when they first examined his body, and he was unable to digest milk as an adult, since he lacked the enzyme that breaks down lactose. In other words, he was lactose intolerant.

These analyses of Ötzi's nuclear DNA confirmed that he was a typical representative of early Stone Age farmers, and that the people of Sardinia are those closest to him today. This can be explained by the fact the earliest farmers spread out over large areas of Europe, including the Alps and

Sardinia. New waves of migration affecting most of Europe have diluted the genetic material inherited from the first farmers. However, more isolated areas on islands and in mountainous areas of southern Europe have maintained clear traces of Europe's first farmers.

CHAPTER TWENTY-FOUR

The Falbygden Area

I have taken the train between Gothenburg and Stockholm so many times before. Mostly I tend to have my nose in a book, but I always look out of the window when the train passes through the area around Falköping.

Nowhere along the way is the rolling, light green, chequered countryside as lovely as it is here. Herds of cattle graze on the hillsides. Spinneys of hardwood deciduous trees appear sporadically among the fields. In the distance loom the flat-topped mountains typical of the region. So it was a pleasant surprise to discover that the earliest known ancestors of my paternal grandmother, Hilda, lived in the Falbygden district. That means that I, too, have some roots in this beautiful area.

Just 10 kilometres (6¼ miles) or so from the parishes where my grandmother's relatives lived in the eighteenth and nineteenth centuries, Swedish researchers have carried out the world's first analyses of nuclear DNA from Stone Age farmers. The parish in question, Gökhem, lies a few kilometres west of Falköping, at the foot of Mösseberg, a plateau mountain. It is home to several of the region's many passage graves, a specific type of megalithic structure built by Stone Age farmers during an intensive period of a few centuries between 5,500 and 5,000 years ago.

The first skeleton in the passage grave at Frälsegården was discovered when a farmer started digging out an earth cellar. Since then, archaeologists have found the remains of nearly 80 individuals. These people appear to have been buried in a squatting position along the walls of the passage graves.

One of them is known as 'Gökhem 4'. Initially, the archaeologists thought the remains were female and referred to 'the Gökhem woman'. However, subsequent DNA analyses have shown that Gökhem 4 was a man. He died aged about 20, some 5,000 years ago. His mitochondrial DNA belongs to haplogroup H, just like that of my paternal grandmother, Hilda. After Ötzi from the Alps, Gökhem 4 was the first Stone Age farmer in the whole world to have his nuclear DNA – not just his mitochondrial DNA – published. Technically speaking, nuclear DNA is far harder to analyse, but it provides far more comprehensive information.

As I mentioned earlier, staff from *Science* came to Uppsala in 2012 to hold a press conference together with Uppsala University. The reason why 5,000-year-old DNA from the Falbygden area aroused so much international interest is that it has a bearing on the question archaeologists have pondered for a century: was it immigrants who spread early farming? Or was the technology itself simply adopted by local hunting populations?

The key point is that the province of Västergötland lies at the very northernmost extreme of the region to which early farming spread. If traces of immigration from the Middle East could be demonstrated so very far north, the question would be settled once and for all. And it has been – thanks to Gökhem 4.

The answer to this old controversy is that agriculture was disseminated largely by immigrants. Several studies of mitochondrial DNA from Sweden and other parts of Europe had already pointed in that direction. However, the new results from Gökhem were based on nuclear DNA, and therefore constituted far more solid proof.

The analyses showed that much of the DNA of the man from the Falbygden passage grave is typical of today's Turkey and Middle East. He was also clearly related to Ötzi, the Alpine man whose nuclear DNA had been published just a few months previously in *Science*'s rival, *Nature Communications*. Conversely, the researchers were able to demonstrate a clear genetic difference between Gökhem 4 and a number of roughly contemporaneous hunter-gatherers from Gotland.

These results aroused strong reactions. Many people, both in the research community and among the general public, were enthusiastic. They could see that new DNA technology had the potential to settle longstanding issues. Others, however, were all the more critical.

Some of these critics can be dismissed as disappointed racists. They disliked the idea that immigrants from the Middle East had played a key role in the early history of the Swedish people. Opinions of this type flourished on blogs and homepages, and I myself received quite a few emails with racist content when I reported the results in *Dagens Nyheter.*

Other critics may be characterised as disappointed archaeologists. Theories that hunters went over to farming independently were an integral part of their lives' work, which now risked being shot down in flames.

'I'm absolutely furious at that Skoglund man,' said one elderly professor to me in the course of an interview.

'I take a critical view of that DNA stuff,' was the first remark addressed to me by another professor, newly retired.

'DNA research has contributed nothing of any value to archaeology,' commented a third professor emeritus.

The arguments rained down. There was some substance and a measure of objectivity in some, less so in others. Particular targets were the young doctoral student and biologist Pontus Skoglund, cited as the first author, and Anders Götherström, who has a background in both archaeology and molecular biology.

Despite the intensive criticism they received from other archaeologists, the researchers who took part in the Uppsala press conference won major accolades after their noted *Science* study. They have published a series of other articles in the world's foremost scientific journals. Three of them – Mattias Jakobsson, Anders Götherström and Jan Storå – have been awarded exceptionally large grants by several of the organisations that finance research in Sweden. The idea is that this finance will enable them to pursue a major DNA mapping project, to produce a kind of atlas of Sweden's first inhabitants.

These research grants are far more generous than those normally awarded for archaeological research. I cannot help but suspect that much of the harsh criticism voiced by other archaeologists springs from envy. No doubt it also partly reflects an inability to come to terms with the new DNA technology and to break with long-established ways of thinking.

'One thing's clear – this criticism has nothing to do with advancing understanding. The motives behind it are very different ones,' comments the German archaeologist Detlef Gronenborn when I meet him in Mainz. He has stopped working in the field of DNA-based archaeology, so tired is he of all the personal antagonism and biased attacks.

The more reasonable criticisms were based on the fact that the research team had only analysed the DNA of one farmer. And he lived 5,000 years ago, when agriculture had already been established in Sweden for nearly 1,000 years. The lack of DNA analyses from the farmers' first few hundred years in Sweden was undoubtedly a weak point. The researchers had not succeeded in analysing the very oldest farmers from Skåne, as the skeletons were too badly decomposed. Conditions in Gökhem have been particularly favourable, partly because the Falbygden area's calcium-rich soils have preserved the skeletons in good condition.

What makes the result so significant nonetheless is the fact that the Gökhem man and the other individuals in the passage grave belonged to the same Funnel Beaker culture that had predominated in Sweden since agriculture first appeared there. If genetic material from the Middle East was still present among Funnel Beaker people in Västergötland 5,000 years ago, it is reasonable to believe that the proportion of such material was at least as significant 6,000 years ago.

Two years later, in April 2014, the Swedish researchers published a new study in *Science*. Now they had compiled analyses of nuclear DNA from four Gökhem farmers, all of whom lived approximately 5,000 years ago. Their DNA was compared with that of seven hunters from Gotland, six of whom were contemporaneous with the farmers from Västergötland, while the seventh lived about 7,500 years ago.

Moreover, the researchers were now able to make comparisons with nuclear DNA from another dozen Stone Age individuals from Europe and Siberia – both hunters and farmers – that had been published by other researchers. Many more sets of DNA from both hunters and farmers from the Stone Age have since been analysed.

The conclusion is plain. The DNA of the early Swedish farmers clearly shows input from the Middle East, Turkey, Sardinia and Cyprus. These farmers differ very clearly both from the earlier hunting people and from hunters living on Gotland during the same period. The widest genetic gap between two populations in today's Europe is that between Finns and Italians. The farmers from Västergötland and the hunters living on Gotland 5,000 years ago were genetically far more different from each other than modern Finns and Italians.

One discernable feature that emerges from DNA analysis is that hunters and farmers probably had slightly different skin colours. Some of the hunters were probably quite dark-skinned. They bore a gene variant common in Africa. The farmers, on the other hand, bore the gene variant which predominates in today's Europe and which is linked to light skin. They also seem to have had brown eyes. It is very likely that dark hair predominated in both groups. But more hunters bore gene variants that indicate they had blue eyes.

Doubtless the two groups also differed considerably in the leather clothing and personal adornments they wore and in the way they embellished themselves with pigments.

Genetically speaking, the hunters from Gotland were not particularly close to the people of any country in modern Europe, nor were they like today's Sami. However, it is clear that they are close relatives of other Stone Age hunters found in Motala, Spain, Luxembourg and Germany. There are even a good many similarities with the 20,000-year-old Mal'ta boy from Siberia.

The hunters can definitely not be described as inbred in the way some Neanderthal people were. They took care to have children with a partner who came from sufficiently far away. Yet they nonetheless show limited genetic variation. This indicates that Europe's original population of hunters was quite small. These people were exposed to harsh privations and to disasters with a high death toll, such as those caused by climate change and flooding.

There was more variation in the farmers' genetic make-up. An explanation for this is that they belonged to a far larger group of people. DNA analyses conducted by Svante Pääbo's team indicate that hunters' numbers multiplied between 15,000 and 10,000 years ago; that is, just when the Ice Age was coming to an end and the climate was becoming milder. But then, between 10,000 and 5,000 years ago, the number of people with the mitochondria typical of hunters declined.

It is also possible to see how the farmers increasingly intermingled with local hunters the further north they went within Europe. Ötzi, who lived 5,300 years ago, was already a descendant of farmers who had come to Europe from the Middle East and local hunters. The proportion of DNA from local hunters was even higher in the Gökhem farmers who lived 5,000 years ago.

Clearly, hunters sometimes moved in among the newly arrived farmers. As a result, children with genetic material from both groups were born in farming settlements.

A pattern can be discerned, supported by Joachim Burger's German study of mitochondrial DNA and isotopes, by Lasse Sørensen's extensive compilation of traditional archaeological finds and radiocarbon datings, and by the DNA results from Sweden and elsewhere. For over 1,000 years, two types of society lived alongside each other, in parallel. The hunters were predominant along the coasts, while the farmers predominated inland.

***

That is exactly what the 'prehistoric village' of Ekehagen shows, albeit on a miniature scale. The settlements are separated by a few hundred metres, rather than a few kilometres or more. Along the shoreline of a small lake stand the crescent-shaped huts of the hunters, while the Stone Age farmers' rectangular longhouses have been built higher up.

Ekehagen, which lies a few dozen kilometres to the south of Falköping, is an ambitious project supported by the local authorities. Buildings from different eras have been constructed on the basis of the best archaeological expertise available. Animals, plants, boats and gear show what life was like in prehistoric times. If you really want to soak up the atmosphere, you can spend the night in one of the houses.

There are traces of Stone Age hunters in the Falbygden region, but they predate the Gökhem skeletons by a long time. The very oldest finds, which come from Lake Hornborga, are about 10,000 years old. Archaeologists have found the oldest remains of dogs in Sweden, including a nearly complete skeleton that looks like a sturdier version of today's Swedish vallhund. However, no traces of hunters' huts have yet been found in the region.

Instead, Ekehagen's huts are inspired by examples from Denmark. To be precise, they are based on Ulkestrup, Zealand, where a Stone Age dwelling has been preserved exceptionally well by being buried in oxygen-free mud.

I peer into the two reed-thatched hunters' huts at Ekehagen. Both of them comprise an inner, oval room and a fairly large covered front yard looking out over the water. On the shore among the huts is a fireplace. There is fishing and hunting gear hanging up, including a fish trap woven from hazel switches and a fishing net made of plant fibres.

At the water's edge lie a few boats made of hollow tree trunks. A few tourists are paddling around in two of them. Wearing brightly coloured lifejackets, they paddle unsteadily but determinedly, splashing loudly. I imagine the hunters of the Stone Age manoeuvred their dugout boats more quietly, with better balance and greater skill.

The summer's day when I visit Ekehagen is unusually cool. But there is a cosy fire inside the Stone Age farmers' longhouse. There is no chimney. Instead, the smoke finds its way along the ceiling, then out through small apertures at the top on the short sides of the house. Although these draw quite effectively, the room is nonetheless smoky. Spending a lot of time in a dwelling like this would not have been good for people's airways. Smoke from simple fireplaces inside homes remains one of the main causes of illness and death in poor countries. There is a good deal of evidence that Stone Age farmers were adversely affected. Ötzi had traces of soot in his lungs, as did several of the individuals from the earlier Çatalhöyük site.

But the heat from the fire is pleasant on this chilly day, and it helps to provide light in the dim room. The only other sources of light are the small holes in the roof where the smoke escapes. A few visiting children, suffering from the cold, huddle around the fire with their parents to eat their packed lunches – sandwiches, presumably.

Cultivated grain – specifically, various kinds of wheat and barley – was also a staple in the diet of the Falbygden area's Neolithic people. They sometimes ate pork, and occasionally there was mutton or beef. However, cows were probably kept mainly for their milk.

It is unlikely that Sweden's first farmers habitually drank milk. A large proportion of Sweden's modern-day population have a particular gene variant that enables us to digest milk as adults. However, this variant is unusual in most of the world. Most adult humans suffer from diarrhoea and severe stomach aches if they drink a significant amount of milk.

Among the original hunting population of Europe, no one seems to have been genetically equipped to drink milk as an adult. Nor does the particular gene variant required appear to have occurred among the earliest population of farmers. Ötzi was unable to digest milk. The same probably applied to the Stone Age farmers from the Falbygden area whose DNA was analysed. It was not until later, at the start of the Bronze Age, that the special gene variant occurred, according to the odd pieces of evidence researchers have discovered. But even then, it does not seem to have been particularly common. It was not a generalised phenomenon until shortly before the Iron Age. Up to that point, most adults were what we today call lactose intolerant.

Yet there is no doubt that the first Stone Age farmers in Sweden kept cattle. Apart from all the ancient bones from cows and oxen, chemical analyses of terracotta vessels have revealed traces of milk fat. The oldest in Sweden are nearly 6,000 years old and come from Fellingsbro in Västmanland. In Poland, ceramic colanders have been found that show people were producing cheese as early as 7,000 years ago.

If I were to speculate, I would say the children living in the Stone Age villages in the Falbygden area drank fresh milk, while the adults stuck to cheese and butter, and possibly yogurt or soured milk. These foods can often be digested by people with lactose intolerance.

The farmers' longhouses at Ekehagen have roofs thatched with reed, just like the hunters' huts. But the longhouses are far more solidly built. Each is supported by a row of large oak posts running down the middle. The outer walls are built of aspen wood.

There are a number of simple millstones in the yard. In each case there is a fairly sizeable stone with a dip in it and an upper stone that you have to move back and forth over the lower one. There are also a few tiny fields surrounded by wooden palisades, where the visitor can try hoeing weeds and clearing ground with only a hand-held flint tool. I can confirm this is pretty hard on the back and shoulders. It is easy to understand how oxen became a welcome addition, once people had properly understood how to harness their strength for farming.

The limestone-based soils in the Falbygden region have some of the best conditions for growing crops anywhere in Sweden. Sweden's first farmers must have noticed this early on. They probably came to the region roughly 6,000 years ago, immediately after making landfall on the shores of Skåne and Halland. They may have followed the River Ätran from the coast and later struck out northwards. It is very hard to discern any signs of the first centuries of agriculture in the Falbygden area. Archaeologists have found some pollen from ribwort plantain (*Plantago lanceolata*), suggesting that the deciduous forests shrank and were replaced by a more open landscape.

But then, some 5,500 years ago, the people of the Falbygden area began building passage graves. Even today, you can hardly fail to notice these burial places, composed of huge stones or megaliths erected here and there in the countryside. In the course of a few hundred years, nearly 300 of these structures were built in the Falbygden area – more than anywhere else in Sweden.

Linnaeus, who travelled through the area in 1747, was among those who described the passage graves. In his book *Västgötaresa* ('Travels in Västgötaland'), he writes of what he calls 'ancestral sites' as 'hillocks raised above ground level, ringed in by stones, covered with earth and grass'.

Similar collections of megaliths have been found from Portugal all the way along the Atlantic coast, in Denmark, Germany and at many places in farming areas of southern Sweden. They are generally regarded as mysterious. However, the US archaeologist David Anthony has an explanation that seems reasonable to me. In his view, the societies that built these barrows were quite simply celebrating their own ability to move heavy objects. Being able to move things was so important for life in an early farming society. Lengths of timber for building houses needed to be moved; grain had to be moved from the fields.

Initially, the only way to move heavy objects was to get large numbers of people to heave on ropes. That called for a great deal of organisation and cooperation. It would also doubtless have required good food and drink – beer, perhaps – to motivate those who contributed their labour. This is what a number of archaeologists believe about Göbekli Tepe, the place of worship in Turkey used by groups of hunters at the dawn of agriculture.

Gathering together megaliths was a way of demonstrating how well organised and powerful you were as a group. Since people had begun to use oxen as draught animals, even relatively small groups of farmers – individual families, indeed – had had the capacity to erect stone structures that would impress anyone who came anywhere near them. A family tomb built of megaliths was the ultimate status symbol. In my opinion, David Antony – and Bernd Zich from Halle – are on the right track here.

While in the Falbygden area, I meet the archaeologist Karl-Göran Sjögren, Sweden's foremost expert on passage graves. It was he who discovered the Gökhem skeletons whose DNA has since been analysed.

Sjögren is a helpful, affable man in his sixties. He picks me up from the station by car and is happy to recount details of his own and others' excavations. However, he is more reserved when it comes to interpreting the finds. His personal interest is in how people in the Falbygden area lived for a specific, limited period. Broad-brush descriptions of Europe as a whole, and surveying several millennia, are not his thing.

He is cautious about drawing conclusions from new DNA results, even though he is listed among the co-authors of one of the groundbreaking publications about the Gökhem farmers that appeared in *Science*. He uses the term 'migrationism', which implies a pejorative view of the idea that inward migration has played a key role in history. Yet he cannot deny the findings set out in the *Science* article of which he is a co-signatory: that the Gökhem farmers whose DNA was analysed were largely descended from people from the Middle East, and that they were clearly genetically different from contemporary groups of hunters living on Gotland. 'Yes, a pattern is beginning to emerge,' he concedes, almost reluctantly.

Despite his sceptical attitude, Sjögren seems to approve of Bernd Zich's ideas that the wooden plough, draught oxen, the wheel and the cart may have developed as the Funnel

Beaker people left the loess soils of central Europe and migrated northwards towards the tougher soils of Denmark and Sweden. 'That would explain a great deal if it were true,' he says. He also agrees that oxen must have played an important role in enabling the farmers in the Falbygden area to assemble the stones for their megalithic tombs.

Even with the help of oxen, building the passage graves must have been a demanding task calling for a good deal of time and resources. Calculations show that 10 men working with a team of oxen needed about a month to build such a grave. Firstly, they had to drag quantities of large stones to the site and raise them to form the chamber itself, which could be up to 17 metres (56 feet) in length. The narrow passageway, which usually faces eastward in the Falbygden area, also had walls formed of large stones. Then an even larger stone, known as the keystone, had to be laid on top to close the whole structure. It could weigh up to 20 tonnes (44,000 pounds).

To enable the oxen to pull the keystone and the other stones forming the roof into position, extensive earthworks first had to be built around the whole structure. The final result was a large, round tumulus with a flat roof. This provided sound protection for the grave chamber. From the outside, the tumulus resembled a smaller version of the nearby plateau mountains, such as Mösseberg and Billingen. Some archaeologists believe the whole point was to create a strong link between the landscape and the passage graves.

Sjögren believes there was probably a degree of social differentiation between farming families in the Falbygden area. Clearly, some families were able to scrape together enough resources to build large, impressive passage graves.

At the time of my visit, Sjögren is leading a major excavation in Karleby, a few kilometres west of Falköping. Some of the very largest passage graves are to be found here, as well as traces of farmers who lived at the same time as the Gökhem people. A large group of students and volunteers are working on the dig, accompanied by a few professional archaeologists from Gothenburg University. They scoop out earth trowelful by trowelful from precisely demarcated squares.

Sjögren shows me the summer's most significant finds, mostly fragments of animal bone, flint tools and pottery. The most notable discovery is a set of holes that once held posts, probably belonging to two houses. He is rather taken aback by the fact that they seem to have been circular; round houses – or, more probably, huts – tend to be associated with groups of hunters. But the people who were laid to rest 5,000 years ago in the Karleby passage grave were definitely farmers. Isotopes show that they lived chiefly on plants, probably mainly wheat and barley. When they ate meat, it was mostly pork from domestic pigs, with some beef and mutton. This is known from bones found at the site. The people who lived here seem not to have eaten very much game, and they ate hardly any fish.

In fact, there are several examples of villages built by Sweden's earliest farmers that feature both round and rectangular buildings; one example is the site excavated at the ESS nuclear research facility under construction near Lund. Round and rectangular buildings may have fulfilled different functions in early farming settlements. The round buildings may have served as granaries, as in Jordan and Cyprus at the dawn of agriculture several thousand years previously.

Isotopes also show that most of the individuals examined from the graves in the Falbygden area were born locally. However, about a quarter of them came from outside the region, possibly from Skåne or from other parts of Väster-götland. This applies equally to men and women. Many of their cattle also seem to have been brought in from elsewhere. This implies that there were networks of farmers extending for many dozens of kilometres.

One of the focuses of Sjögren's research has been the objects found around the passage graves. The grave chambers themselves often contain amber beads, animals' teeth and small axes. Shards of pottery are usually to be found a few metres in front of the entrance to the chamber. They are a delicate, attractively patterned variant of Funnel Beaker vessels, far more meticulously crafted than the everyday pottery found in the settlements. These attractive vessels next to the passage graves appear to have been smashed on purpose. It is common for archaeologists to find the burnt bones of oxen, pigs, sheep or dogs and burnt fragments of flint in the same place. Curiously enough, however, there is no sign of any fire. The objects seem to have been burned somewhere else and later placed in front of the entrance to the grave chambers.

This pattern, involving the same type of objects, recurs at one tomb after another. It looks as if special ceremonies were held that involved breaking pottery and burning flint axes and animal bones. Other rituals involved dropping axes and amber beads into lakes as votive offerings – and sacrificing living people.

**

There are about 30 known cases in the early farming cultures of northern Europe where people appear to have been sacrificed in lakes. They are generally known as bog bodies or bog people, as the shallow lakes gradually filled up with plant life and developed into peat bogs. Both men and women have been found in such bogs; those who died in the Neolithic period were mostly in their late teens.

The oldest of Sweden's known bog people is the Raspberry Girl, who lived at the same time as the Gökhem farmers and was found a few dozen kilometres further south, in a bog in the parish of Luttra. Today she is an exhibit in the Falbygden Museum.

Visitors to the museum can view her brown skeleton in a display cabinet. It is quite small. The Raspberry Girl was about 19 years old when she died, but only 1.45 metres (4¾ feet) tall – that is, some 10 centimetres (four inches) shorter than the average height of women at that time. She also appears to have been of unusually light build, and her face seems to have been fine-featured. Her teeth were even, complete and well formed, although the chewing surfaces were rather worn.

Immediately before her death, she ate a large quantity of raspberries. Her stomach contents, including the yellow raspberry seeds, are clearly visible alongside the skeleton in the display case. This shows she died in late summer when the raspberries were ripe, in July or August.

Researchers believe she was killed, rather than drowning in the lake by accident, because of the appearance of the skeleton when she was found. The position of her legs reveals that they were bent right back and bound, so that her feet were next to her buttocks. It also looks as if her wrists were bound together. In that position, she was thrown – dead or alive – into a shallow lake. Little by little, the lake filled up with vegetation and turned into a bog.

She was found there in 1943, some 5,000 years later, by a man who was digging up peat for fuel. As soon as it had been established that she was a prehistoric individual, she was sent in a large block of peat to Stockholm's History Museum, where she was examined and presented in a scientific report. After a few decades on display in the History Museum, she was put into storage. It was only in the 1990s that she returned to the Falbygden area, where she is now one of the main attractions at the local museum.

In addition to the actual skeleton, the Raspberry Girl is displayed as a life-size silicone model, a model-maker's impression of how she may have looked. It depicts a short, fine-boned young woman, just like the skeleton itself. Her features have been carefully built up on the basis of her cranial bones, in accordance with standard techniques. That is all just as it should be.

Yet I have some doubts about the Raspberry Girl's hair and skin, and the colour of her eyes. Firstly, I imagine that a 19-year-old woman – even one living in the Neolithic period – would have been more particular about keeping her hair nice-looking and attractively styled. The model's hair, however, is rather straggly and unkempt. It is quite fair, close to honey-blonde, and she has a very pale complexion. She may well have looked like that, of course, but we cannot be sure. If she was one of the farmers, she is more likely to have resembled the people of modern Sardinia and Corsica, with brown eyes and dark hair.

Her fair skin may be an accurate representation – if she came from a farming family. But if they sacrificed a girl from outside their community, from a group of hunters, it is equally probable that her complexion was considerably darker.

CHAPTER TWENTY-FIVE

Hunters' and Farmers' Genes

There are those who believe agriculture to have been the most disastrous invention in history. The American author and physiologist Jared Diamond, for example, once described it as the 'worst mistake in the history of the human race'. It is a common view that diseases, class divisions, war and misery of every other kind befell us when we abandoned our lives as hunters and made the transition to farming-based societies.

Sometimes I am inclined to agree that things were better during the age of the hunters, before we started to farm the land. That would be my view on comparing the Raspberry Girl and the Österöd woman, for instance, two of the most important finds of female individuals from prehistoric Sweden.

On the one hand, we have a 19-year-old slip of a girl who, for the sake of some murky beliefs, was bound and thrown into a lake to drown. Although the nature of these beliefs is obscure, they may have been linked in some way with the harvest and supernatural forces. The Österöd woman, on the other hand, was a well-nourished, athletic 80-year-old with teeth in better condition than those of most Swedish pensioners today, who died in the archipelago off Bohuslän after a long and active life as a hunter.

If I were to swap places with one of them, there's no doubt who I'd choose to be – the Österöd woman. However, we need to bear in mind that she represents a unique stage in Sweden's early history. She and the group she belonged to came to a virginal land, where nature provided more than they could ever need. They had no competitors; they could fish, hunt dolphins and elk, and gather shellfish, berries and nuts to their hearts' content. When the population expanded during the Mesolithic period, life became much harder.

I also ponder the fates of other individuals I have heard about during my journey to prehistoric Europe – such as the young Ice Age woman at Abri Pataud who died with her newborn baby 30,000 years ago. She may have lost her life through giving birth. However, her teeth were so inflamed and so badly damaged that that alone would have been enough to cause a protracted and agonising death.

It is simply not true that we humans first encountered lifestyle diseases through agriculture, nor does there seem to be any decisive difference in infectious diseases before and after the advent of farming. Tuberculosis, for instance, seems to have been with us at least since some of us left Africa over 55,000 years ago. When DNA researchers compare the immune defence of early European hunters with that of farmers several millennia later, they cannot discern any definite signs of an increase in the incidence of disease.

***

One variant of the low-carbohydrate diets so trendy these days is known as the 'Paleo diet' and is supposedly inspired by the Palaeolithic, or Old Stone Age. It involves eating a high proportion of meat and fish and avoiding dairy products and cereals, which only became common with the emergence of farming. Although a Paleo diet includes some sweet berries, there is very little place for starch.

Such a diet may well have some advantages in terms of both flavour and health properties, compared to today's hyper-processed food with its high sugar content and fast carbohydrates. But the advocates of Paleo diets often give an extremely simplified – indeed, mistaken – account of what people ate in Palaeolithic times.

Firstly, people like us, anatomically modern *Homo sapiens*, lived through the Old Stone Age for at least 100,000 years, on all the continents except Antarctica – a long period, in many different locations with disparate conditions. We lived in tropical, temperate, Arctic, dry and wet climates. Our living conditions varied enormously. Some Palaeolithic peoples ate mainly reindeer meat, others almost exclusively seal meat and fish, while still others lived largely on starchy roots. Insects and other small creatures were often an essential part of the diet. Everyone who had access to honey consumed it. Ethnological studies of today's hunter-gatherer peoples show the importance of honey to people everywhere that wild bees are to be found – which is in most places apart from those at Arctic latitudes. Traditional groups such as the peoples of the rainforest and the Hadza of East Africa obtain a considerable proportion of their nutrients from honey, which they collect from several species of bees.

So far, archaeology has given a distorted picture of what people ate in the past, as the remains of large mammals and shellfish are preserved so much better than those of plants. Bones and oyster shells are visible evidence, even after several millennia. However, new technology is now emerging that can reveal the part played by plants in the diet. For instance, it is now possible to examine microscopic deposits in the tartar on human teeth. One of the leaders in this field is Amanda Henry, a colleague of Svante Pääbo at the Max Planck Institute for Evolutionary Anthropology in Leipzig.

She examines the surfaces of ancient teeth for microscopic traces of starch and minute traces of silica from different types of plant, which differ in appearance according to species. Analyses carried out by Henry and her collaborators show clearly that people have been eating plant material for a very long time. People in the caves at Blombos in South Africa and Skhūl in Israel were already eating different types of grass seed 100,000 years ago. The people of Skhūl also had traces of dates on their teeth. During the Ice Age in Europe, people ate grass seeds and the starchy roots of plants including water lilies and bulrushes. This has been shown by analysing teeth from sites such as Cro-Magnon, Abri Pataud, Dolní Věstonice and La Madeleine.

Simple mortars in which microscopic traces of starch have been preserved have also been found at sites including Kostenki (Russia), Biancino (Italy) and Pavlov in the Czech

Republic. All of these date back about 30,000 years. The starch seems to have come mainly from the roots of bulrushes and ferns.

Amanda Henry says she is irritated by those who seek to make money by writing books claiming that people in general have an 'evolutionary adaptation' to a particular way of eating. There is no scientific proof that any particular kind of 'Paleo diet' suits everyone best.

In other words, the notion that evolution has shaped us to do without carbohydrates almost completely is mistaken, even though some pundits want to give this impression. We are genetically equipped, to a far greater degree than apes, to digest starch in our food. The reason for this is that starch in the diet has played an important part in our survival for the last six million years, since our evolutionary branch diverged from those of our nearest relatives, the chimpanzees and bonobos.

However, we may differ in our sensitivity to the large proportion of starch typical of the modern diet, and our genetic background may account in part for that sensitivity. After all, there can be no doubt that the switch to agriculture represented a dramatic change. Most farming societies throughout history have eaten more starch than most hunting societies. Even our dogs changed their way of life, as is shown by a number of interesting DNA studies.

Kerstin Lindblad-Toh, a Swedish researcher based in Uppsala, is the world's leading expert in canine DNA. Comparing the genetic material of various breeds of dog – which are often exactingly bred to meet pedigree specifications – is an ingenious shortcut to identifying disease-bearing genes that also occur in human beings.

In 2012, Lindblad-Toh and her collaborators made a contribution to the heated debate on when dogs were domesticated. The main conclusion their study reached is probably incorrect. The team suggested that wolves became humans' companions only when we took to farming – that is, slightly less than 12,000 years ago. This cannot be right, as there is convincing archaeological evidence – and other DNA studies – that show dogs must have been with us for several millennia before that.

However, the results of the study are very interesting anyway, as regards both the history of dogs and that of humankind. This is because the Uppsala researchers managed to show that dogs, just like people, have a number of genes whose function is to break down starch. These genes emerged after dogs had started living together with humans; wolves have them to a far lesser degree. The conclusion is thus that dogs have adapted through evolution to the food they have been given by farmers over the millennia, which includes quite a large proportion of leftover porridge and bread.

The one minor error in this initial conclusion is that it is not true of absolutely all dogs. A few years later, an international team of researchers published a new study. They had examined the DNA of more breeds of dog, including some very old ones. Their study showed that a number of ancient breeds, such as dingoes, basenji and Siberian huskies, resemble wolves in that they are not genetically equipped to break down starch. Lindblad-Toh's team subsequently confirmed these findings and added further old breeds, such as Greenland sled dogs.

So the conclusion to be drawn from these latest results is that some dogs were already living with humans when we were hunters, and were genetically adapted at that time to live almost exclusively on meat. Dogs adapted to consuming farmed food at a later stage, when their human companions went over to farming. A parallel can be drawn here between dogs and people. Farmers have largely displaced Europe's earlier population of hunters. At the same time, farmers' starch-eating dogs have multiplied at the expense of the older breeds of dogs that accompanied the hunters.

Humans, too, have undergone a similar genetic adaptation to farmed food. For example, there is a gene known as AMY-1 whose function is to break down starch by means of saliva in the mouth. Our cousins the chimpanzees normally have a single gene that works like this. Bonobos, which are equally closely related to us, have none. A human being, on the other hand, can have several copies of the AMY-1 gene. Some have only a few, while others can have up to 15 or 20 copies. Individuals who have many copies can break down starch more efficiently.

And the capacity to break down starch efficiently has been vital for survival in certain human environments, but not in others. This is why the Hadza, a people of hunter-gatherers living in East Africa, have a comparatively large number of copies of AMY-1. They live in a dry tropical climate and live largely on starchy roots that they dig up out of the ground. The Aka and Mbuti peoples also live in the tropics, but in regions with humid rainforests. Their traditional diet contains more animal protein and sugars from fruit and honey, and they tend to have fewer AMY-1 genes. People from farming cultures based on the cultivation of grain, such as the Japanese, have significantly more genes of this type.

It is simply the case that the capacity to break down starch varies depending on the type of environment in which our ancestors lived. Their lives – whether they were hunters or farmers – marked our genes in different ways.

Most people living in today's Europe have a mix of 'hunter genes' and 'farmer genes'. We represent different waves of immigration, which affected the early history of our continent in different ways.

Take me, for example. Following my own mitochondria and thereby my direct maternal lineage, I can go back several tens of thousands of years to Ice Age forebears who hunted reindeer and other deer. Following my paternal grandmother's mitochondria, my maternal lineage goes back to Syria and the cradle of early agriculture. I am a mix. We all are.

However, the fact is that the first wave of 'hunters' and the second wave of 'farmers' do not really account adequately for the early history of Europe. DNA researchers are beginning to be increasingly convinced that there was a third wave.

That wave came from the east.

PART THREE

THE INDO-EUROPEANS

Måns was the father of Per, who became the father of Johan.

Johan's son was Per, and Per became the father of Johan.

Johan had a son called Nils, who had a son called Peter.

Peter's son was called August.

August became the father of Eric, who became the father of Göran, Anders and Gunnel.

Göran had a daughter, Karin.

CHAPTER TWENTY-SIX

The First Stallion

The lowing of cows and oxen was to be heard on the plots of the Gökhem farmers in Västergötland 5,500 years ago, along with the bleating of sheep, the barking of dogs and, in the woods around the farms, the grunting of pigs. At the same time, in the steppe far to the east, people were beginning to make use of an entirely new domesticated animal.

Of one thing we can be sure: there were domesticated horses in the region we now call Kazakhstan about 5,500 years ago. The oldest incontrovertible proof that has been found comes from the Botai culture.

The Botai people lived in the steppes north-east of the Caspian Sea. They were primarily hunters, though comparatively sedentary, and their diet consisted largely of horse meat. Over 99 per cent of all the bones found in their settlements come from horses. The Botai people rode their domesticated horses and hunted wild ones. Many of the bones from their settlements come from particularly fine-boned or gracile animals, which researchers view as an indication that they were domesticated.

There is even better evidence that some of the horses at the time of the Botai culture were domesticated. One piece of evidence is milk. Traces of fat on terracotta shards show that the pots from which they came were used to store milk. Even today, horse meat and foodstuffs made from mares' milk are part of the traditional diet in Kazakhstan and the neighbouring steppe regions.

Damage to horses' vertebrae suggests that the animals concerned carried riders. And there is even clearer proof in the marks on the tooth enamel of some horses, which were clearly caused by bridle bits. The bridles were very probably made of leather and wood, and most of the stone tools from the Botai culture seem to be designed for leather-working.

There is thus a consensus that horses were domesticated at least 5,500 years ago, and that there were domesticated horses in the northern steppes of the region that is now Kazakhstan. However, the fact that the oldest undisputed archaeological evidence comes from the Botai culture does not necessarily signify that these people were the first to domesticate horses. There may have been domesticated horses elsewhere at an even earlier stage, though the evidence is less conclusive.

DNA researchers have also sought to identify horses' origins. The focus has been mainly on examining living horses of all possible breeds. The first large-scale analyses, which were carried out in Uppsala, revealed an interesting pattern. Modern horses have a multiplicity of different mitochondrial lineages. The only possible explanation for this is that people in many different places captured wild mares and used them for breeding. In contrast, only one Y-chromosome lineage was identified. That means that wild stallions were domesticated within one area only, and that there were only a few such cases – possibly just one, in fact.

These patterns make sense if we reflect on how wild horses live in their herds. Stallions are far more aggressive and independent. Mares, on the other hand, are used to living in a group and to accepting subordination to a higher-status horse. They are typical herd animals, just as cows are. It is a fairly straightforward matter to tame a wild mare; taming a wild stallion, on the other hand, is all but impossible.

I imagine the first domesticated stallion may have been a little foal that followed his mother when she was captured by humans. The youngster must have been unusually gentle-natured. He would undoubtedly have been outcompeted by other males in the wild, and would never have had any offspring of his own. Under people's care, however, he was able to grow to adulthood and serve the mares of the herd, thereby becoming, several millennia later, the ancestral father of all the horses living today.

Unfortunately, DNA research has not yet been able to provide any more precise information about where the first stallion was caught and where people first started to breed horses. The evidence points towards a fairly extensive region stretching from western Kazakhstan through the Russian steppe to the Volga and the Don, and possibly towards Ukraine.

One reasonable scenario is that farmers with domesticated animals spread from modern Turkey to the regions north of the Black Sea. On reaching the steppe, they faced a number of problems. There were endless amounts of excellent pasture for half the year. But the winters were harsher than they were accustomed to. Their sheep could not cope with a frozen crust of snow, and even the thinnest layer of snow prevented the cattle from grazing.

While the farmers' own animals were dying of cold, they could watch herds of wild horses gracefully scraping the snow from the pasture they needed with their hooves. When the horses needed to drink, they used their hooves to make holes in the layer of ice covering springs and streams. These horses were well adapted to living in a climate with cold winters; that was how they had got through the Ice Age so successfully in the European steppes. They were not suited to living in forests, however. When the forests emerged after the end of the Ice Age, most wild horses disappeared from Europe. It was only in the steppes in the eastern part of the continent that they were still to be found in large numbers.

Those who found ways of capturing and controlling horses benefited from a domesticated animal that was largely self-sufficient and provided an excellent reserve of meat during the cold winters of the steppe. And the step from providing a meat reserve to being a dairy animal was a short one.

People who are used to milking cows can learn to milk mares too. And just think what a bonus it must have been when people discovered the particular virtues of mares' milk. It contains a good deal more lactose than cows' milk and is unsuitable for drinking when fresh, particularly for people with lactose intolerance – nearly all adults at that time. But those who drank mares' milk that had been kept for a while must have felt a little bit warmer, more relaxed and more talkative. The high sugar content of horses' milk makes it ferment easily, producing a small percentage of alcohol.

***

The Samara Valley, near the River Don in Russia, was the site of the Samara culture. In the valley archaeologists have found the remains of horses over a thousand years older than those of the Botai people. The remains of 10 individuals have been found at an ancient burial site. They were buried with gifts such as red ochre, broken pottery and beads made from seashells. Archaeologists have also found the heads, hooves and lower legs of two horses in the immediate vicinity of the graves. Close by lay two small horse figurines made of bone.

One interpretation is that the people who took part in the burial feasted on the horse meat, after which they laid the horses' hides, together with their heads and bones, in the graves of the dead.

The Samara people lived in the northern part of the steppe, near the margins of the forests. It is quite clear that they kept domesticated cattle and sheep, and, to judge from finds of bones in their settlements, they also hunted deer and beaver. This meant their diet included meat from both domesticated and wild animals. Many, though not all, experts are convinced that the horses of the Samara culture must have been domesticated and that they were used at the very least as a source of meat. They may also have provided milk and been ridden by the Samara people, but there is some disagreement on this.

The US archaeologist David Anthony has made tremendous efforts to locate the oldest evidence of riding. The items he and his wife and collaborator Dorcas Brown have collected include a large number of horse teeth – some of which bear the traces of a bridle bit, while others do not – which they have examined under a scanning electron microscope. They have also ridden horses using different types of bit that they believe early riders would have used. They have tested bits made of leather, horsehair, woven hemp and bone. After 150 hours of riding, they made models of horses' teeth at the point where the bit had rested, and studied the damage to the teeth under the microscope. They then looked for similar damage to horses' teeth that they and other archaeologists had found in excavations in the steppe.

David Anthony also draws attention to the regions around the Don and the Volga. He makes particular mention of the herding culture known as the Khvalynsk culture, which occupied approximately the same area as the Samara culture, though slightly later. These two cultures, and others, gave rise to the Yamnaya (Yamna) culture, which would eventually dominate large areas of the steppe – all the way to Ukraine in the west. It is clear that horses were important to these people. This is shown by finds in their typical tombs, known as kurgans. The kurgans consisted of a pit, often lined with timber, which was covered with a tumulus made of stones and peat. There is much evidence that horses were sacrificed at such burials, as were cattle, sheep and dogs.

The Yamnaya culture took off 5,300 years ago, at exactly the same time as the farmers in the Falbygden area were most active in building megalithic passage graves. There are three factors that may explain why the Yamnaya culture became so successful and extended over such a vast area.

Firstly, the climate changed at that time. The steppes became colder and drier. Pollen analyses show that tree cover declined, while plants well adapted to a dry climate, such as wormwood (sagebrush), became more common. The steppe had never been a particularly favourable environment for cultivation, as the soil was often too dry, sometimes too marshy, and not infrequently too salty. But the spacious grasslands were ideal for animal husbandry, especially for keeping flocks of sheep, which are better adapted to a dry climate than cows.

Secondly, the Yamnaya herdsmen on the steppe had a vital factor in common with the Funnel Beaker farmers of the Falbygden area: they were beginning to use oxen as draught animals, along with wheeled vehicles. We know this because both wheels and vestiges of carts or wagons have been found in the typical kurgan graves of the steppe.

Thirdly, they were probably beginning to master the art of riding.

The combination of the drier climate, the ox-drawn wagons and horses that could be ridden brought dramatically altered conditions for animal husbandry. The Yamnaya culture, which focused more on herding animals than on tilling the soil, drew the winning straw. People could pack everything they needed into their ox-drawn wagons, including tents to spend the night in and drinking water for at least a week. This enabled them to live in the steppes for months and drive their animals on to fresh pastures. When an area had been grazed to the limit, all they had to do was lead their animals onward to new green meadows. With the help of horses, they were able to control many more cattle and sheep than before.

Before the time of horses that could be ridden, oxen and wagons, the gigantic steppe regions between the river valleys were like a vast ocean to people without seaworthy boats – infertile, impenetrable land. With the help of horses and wagons, however, the Yamnaya herdsmen could take possession of the whole steppe. Their herds grew larger and larger, while they grew richer and richer. They acquired more and more *fä*, to use an old Swedish word of ancient Indo-European origin for livestock. The word for livestock (*peku* in reconstructed Proto-Indo-European, *pecu* in Latin) acquired an extended meaning in the colloquial Swedish *pekiner*, which became *penningar* and finally *pengar*, the Swedish word for money (English 'pecuniary' comes from the same root).

David Anthony has summarised his ideas in a 568-page book entitled *The Horse, the Wheel, and Language*. In this work he attempts to piece together all the evidence that suggests that the Yamnaya culture, with the help of the horse, spread the Indo-European languages in all directions.

And it was not just the Indo-European languages – now including English, Swedish, German, Spanish, Russian, Persian (Farsi), Pashto and Hindi, spoken today by about half the world's population – that were disseminated. The same applies to a whole package of innovations, including the alloy bronze, woven woollen cloth, long-distance trade, new gods, new rules on inheritance and justice, wider class divisions, worse wars, increased power for men, and more.

In short, the very society we still live in today.

***

David Anthony is just one in a long line of researchers to make connections between the Indo-European languages and the horse. Another of the better-known ones was Marija Gimbutas. Although not the first either, she is generally considered to be the great pioneer of the steppe theory on the origins of the Indo-European languages.

Lithuanian-born Marija Gimbutas fled via Austria to Germany when the Soviet Union invaded Lithuania during the Second World War. By her own account, she left holding the hand of her little daughter in one hand and her diploma in the other. In 1946 she received a doctorate in archaeology at Tübingen University. Three years later she moved to the US, and after a tough start – she faced discrimination as a woman and a foreigner, and had three small children – she eventually forged a long and successful career as an archaeologist. She became a professor of archaeology at the University of California, Los Angeles (UCLA) and led excavations in Greece and the Balkans for many years. She wrote a number of books that were well received in some quarters, not least by a growing feminist movement. Jean M. Auel, for instance, the author of the bestselling *Earth's Children* series of books, was influenced by her theories. However, Gimbutas also met with strong resistance from a number of her peers.

Among the basic components of her theories is the idea that hunters and early farmers in Europe up to about 5,000 years ago were essentially egalitarian and peaceful. Their religious life was dominated by belief in a powerful goddess, and their societies were matrilineal, meaning that property, such as houses and land, was passed on from mother to daughter.

According to Gimbutas, these peaceful matrilineal societies came to an abrupt end when they were crushed by warlike patriarchal tribes that came riding in from the steppes in the east. The men on horseback brought with them Indo-European languages, which they imposed on everyone; the package also included Indo-European religion and mythology.

No one today would deny Gimbutas's encyclopaedic knowledge of archaeological sites and finds reflecting Europe's early farming cultures. What is more controversial is the notion that 'Old Europe', as she called these early farming cultures, was composed of peace-loving matrilineal societies.

She also interpreted material largely in the light of a great omnipotent goddess, whose presence she detected in every conceivable context. She believed she could discern signs of such a goddess cult everywhere: on pottery decorated with zigzag patterns, in cave art depicting circles crossed by a line and in rounded buildings. As regards the circles crossed by a line, which feature in a number of Ice Age cave paintings, many of today's researchers share her view that they may have symbolised female genitalia. However, there is quite a step from that to an omnipotent goddess. And the idea that round buildings had a similar meaning has few advocates these days.

True, large numbers of female figures have been found that may well represent goddesses. I have seen many of them at a major exhibition in London's British Museum and at smaller museums all over Europe. I have seen the 40,000-year-old Venus of Hohle Fels and female figures with bent knees and extended arms from farming settlements in Cyprus that are about 6,000 years old – not to mention hundreds of female figures from the millennia in between. At least two of the experts I have spoken to, Jill Cook of the British Museum and Carole McCartney from Cyprus, are convinced that such figures were linked with birth, fertility and female rites. However, there are also many figures, paintings and decorations with quite different motifs, such as the Lion Man from Hohlenstein-Stadel (also over 40,000 years old), the Ice Age cave paintings of wild animals, and Çatalhöyük's murals depicting leopards. The idea that they, too, symbolise a goddess is more far-fetched.

The dominant view today is that hunting societies and the early farming societies were comparatively egalitarian in terms of social differentiation and gender relations. However, there is no convincing evidence at the present time of any matrilineal systems. This is an issue that could potentially be resolved by analyses of DNA and isotopes. Existing findings suggest that both men and women moved to other settlements. As regards the hunters from Motala, the young women seem to have moved, while the men do not. A man has been found in Hungary whose DNA set is very clearly characteristic of a group of hunters, but who apparently lived in an early farming settlement. About a quarter of the early farmers from the Falbygden area moved in from elsewhere, and they include both men and women. However, far more samples would be needed to be able to discern a pattern – if there actually is one.

Hunters and early farmers cannot have been so very peace-loving. In the Ice Age settlement at Dolní Věstonice, which dates back 30,000 years, it was common for people to hit each other over the head, leaving permanent injuries to the skull. In the Swedish hunters' settlement at Tågerup, Skåne, a young boy died 7,000 years ago when someone shot him in the back with an arrow. Many other Stone Age settlements have yielded skeletons showing injuries due to violence. A number of early farmers in Talheim, southern Germany, were simply massacred 7,000 years ago; it looks as if 34 individuals were killed on the same occasion.

The view today is that some of Gimbutas's interpretations are speculative, while others have been shown to be incorrect. However, archaeologists are beginning to catch up with her in other respects. Throughout her long career – from the Second World War to the 1990s – there were few archaeologists

who thought inward migration had had any great impact
on Europe's development. Such views were referred to,
slightingly, as 'migrationism'. The dominant view was that
human societies have the capacity to develop of their own
accord, and that there is no need for anyone to come from
elsewhere with innovations.

However, as we have seen, new DNA research shows
clearly that immigration was hugely influential on Europe's
development.

**

It is no mere coincidence that Marija Gimbutas came from
Lithuania. She had been brought up on stories about supernatural beings in the forests. Her nursemaids told her about
sacred goddesses, whom they took very seriously. Both her
parents were doctors, and they took an interest in research
into Lithuanian folk culture and history. Traditional musicians
and experts in old folk tales often visited Gimbutas's childhood
home. As an adult, she spoke a number of Indo-European
languages, including German, English and Russian, and she
could understand several Slavonic languages besides Russian.
Her mother tongue, however, was Lithuanian.

Of all the Indo-European languages still spoken today,
Lithuanian is the most archaic. Moreover, Lithuania was one
of the last countries in Europe to adopt Christianity; in
practice, it was not Christianised until the sixteenth century.
And pre-Christian beliefs lingered on among country people
for much longer. In the nineteenth and the early twentieth
century, industrious folk culture researchers went around
Lithuanian villages writing down old folk tales and songs.
This was part and parcel of the national romantic zeitgeist.
Similar material was being collected in many other places,
not least in Germany.

This enthusiasm for Europe's 'original' heritage was far
from unproblematic. It was to become part of the package of
racial doctrines that would have such devastating consequences in the Holocaust. However, one of the positive

outcomes of the folk culture researchers' work was the extensive collection of old tales and folk songs from the whole Indo-European linguistic region that can now be studied by today's researchers.

Contemporary Indo-European languages can also be compared with Old Church Slavonic, ancient Vedic scripts in Sanskrit and texts in other Indo-European languages that are now extinct. This makes it possible to reconstruct how the first Indo-Europeans may have spoken, and even how they lived and what kind of conceptual world they had.

For instance, it seems they had access to oxen, wheels, axles and wagons. 'Wheel' and 'axle' (*hjul* and *axel* in Swedish) are words with ancient Indo-European roots, as is 'yoke' (*ok* in Swedish). They can be found in nearly all branches of the Indo-European language family. 'Ard' (*årder* in Swedish) – the early wooden plough that came into use at about the same time as wheels and wagons – is another example.

Researchers have made huge efforts to puzzle out what animals and plants were to be found in the Indo-European linguistic area before it diverged into separate branches, such as Italic, Germanic, Slavic, Baltic, Celtic and Indo-Iranian. In this way, they are trying to identify the type of natural surroundings where the language first developed. They have found both wild animals, such as beavers, otters, lynxes and bears, and domestic ones, such as oxen, cattle, rams, ewes, lambs, pigs and piglets. And, as you might expect, horses are on the list. Reconstructions suggest the horse was designated by a word similar to *equus*, its Latin name.

It is absolutely clear that the early Indo-Europeans used the wool from their sheep. The words 'wool' and 'weave' (*ull* and *väva* in Swedish) have their origins in Proto-Indo-European.

There is a snag in that both *lax* (Swedish for 'salmon') and *bok* (Swedish for 'beech') are words of ancient Indo-European origin. Yet neither salmon nor beech trees were in evidence in the steppes. However, proponents of the steppe theory try to get round that problem by suggesting that the names may originally have referred to other salmon-like fish living in the rivers and streams of the steppe, and to other types of tree.

It must be more encouraging to them that the old Indo-European word for honey resembled 'mead' (Swedish *mjöd*). The cognate is still in use in countries such as Russia and Poland, both for semi-liquid honey direct from bees and for the fermented honey-based alcoholic drink we know as mead. This beverage seems to have played a key role in the feasts and rituals of the early Indo-Europeans. The fact that wild honey bees are less common in Siberia supports the idea that the first Indo-Europeans came from the European part of the steppe.

Another set of clues that can help in trying to work out where the first Indo-Europeans lived is the loanwords that entered the languages at an early stage. These give us a clue to the identity of their closest neighbours. It so happens that the neighbouring language with most impact on the early Indo-Europeans seems to be an early variant of Finno-Ugric – that is, the forerunner of the languages that would later evolve into Finnish, Estonian, the Sami languages, Hungarian, and others.

Their respective vocabularies suggest that the Finno-Ugric speakers lived in forested areas further north, while the Indo-Europeans lived in the steppe further south. Linguists may also be able to discern the influence of Semitic languages prefiguring modern Arabic, Hebrew and Aramaic.

Most serious experts in recent years have agreed that the Indo-European languages began to spread out from a region somewhere near the Black Sea. Some experts think this began on the south side, in modern Turkey. However, the prevailing view is quite close to that advanced by Marija Gimbutas in her day: that the huge dominance of Indo-European speakers in the world originated somewhere in the steppe, in the region north of the Black and the Caspian Seas.

One of the people who has built on Gimbutas's legacy is her former pupil James Mallory, a US citizen who has been working in Northern Ireland for many years. He is the author of several of the weightier standard works on the origins of the Indo-European languages. For many years now he has been the editor of the leading interdisciplinary journal in the field, *The Journal of Indo-European Studies*.

I travel to Belfast to meet him. Now a professor emeritus, Mallory has been assigned a minuscule study at Queen's University, which he is obliged to share with another retired professor. Both have brought stacks of books with them, which are crammed in along one wall. They have also attempted to create a dividing wall as best they can by placing a bookcase between their respective sections. There is no space in the little room to conduct an interview, so we go down to a ground floor café to talk.

Mallory grew up in California. He studied history and subsequently worked in the military for several years before ending up in UCLA as a doctoral student. Gimbutas was both his mentor and his landlady. He and other postgraduates got to live in a little cottage on her land. They paid a peppercorn rent, and their rental contract provided for them to help out on Gimbutas's plot by weeding and chopping wood.

That was in the early 1970s. Gimbutas was regarded at UCLA as mildly eccentric, with her strong Lithuanian accent and all her amber jewellery. But she was a highly respected archaeologist, and her knowledge was considered to be almost encyclopaedic. Her teaching was solid, but old-fashioned. Lectures consisted largely of rattling off a plethora of details of different archaeological cultures, while the students took notes and copied down patterns and shapes. It provided an excellent foundation, says Mallory, but he is glad he learned from other tutors about modern scientific method: how to set up hypotheses and test their validity through investigation.

It was on a dig in the Balkans that Mallory first heard Gimbutas talk about the omnipotent goddess of whom she believed she could see so many signs. The more emphasis she placed on the more polarised students and other archaeologists became in their views of her. Some saw her almost as the goddess she described, while others increasingly distanced themselves from her.

Mallory himself appears to look back on his old tutor with affection and respect, though also with a certain distance. Her theories about how steppe cultures spread the Indo-European languages have experienced a revival today. However, Gimbutas's claims that earlier farming cultures in Europe were so very peaceful have not stood the test of time. Whether inheritance was essentially matrilineal before the Indo-Europeans entered the picture is an open question.

We sit talking in the café for a few hours, and Mallory is very friendly and open. But he is noticeably uncomfortable with some of my questions when I get on to the subject of how the far right have influenced research into the origins of the Indo-Europeans.

However you look at it, there is no denying the subject has attracted many nationalists, anti-Semites and racists over the years. For example, the owner of *The Journal of Indo-European Studies*, which Mallory edits, is a controversial American called Roger Pearson. While he was listed as a member of the editorial board, some researchers refused to work with the journal. There are a number of reasons why they regard Pearson as politically unacceptable. One is that he set up the Northern League for North European Friendship in the late 1950s. The League's influential members included Nazi Germany's leading 'race biologist', Hans F. Günther.

Roger Pearson is nearly 90 now. He has undergone several heart bypass operations, and he is no longer listed as a member of the *Journal*'s editorial board. Mallory makes it clear to me that he totally disagrees with Pearson's views, such as the supposed existence of races hypothetically linked to different levels of intelligence. However, he believes democracy should allow researchers to write about crackpot theories, including politically sensitive ones. Moreover, if Pearson did not publish the *Journal of Indo-European Studies*, who would? Mallory hopes to see one of Pearson's sons take over soon.

I have never heard anyone claim that Marija Gimbutas herself was a fascist. Yet she worked together with Pearson and other researchers who have been accused of being fascists, anti-Semites and Nazis. And there is a passage in her book *The Language of the Goddess* that gives me pause for thought.

In the final summary, she refers to two periods in the history of Europe that she regards as particularly dark, particularly from the point of view of women. One of them was the Christian Inquisition of mediaeval times, when many women were put to death as witches. The other was Stalin's regime in the Soviet Union, under which millions died. She makes no mention whatever of Nazism, Hitler's Germany or the Holocaust.

When we return to Mallory's study, he opens the book. He has never before noticed that Gimbutas referred in her summary to the Soviet Union, but not to Nazi Germany. After pondering this for a while, he admits it looks peculiar. His conclusion is that Gimbutas was strongly affected by her own experiences as a young woman fleeing the Soviet troops. She was a Lithuanian nationalist, and her homeland was occupied by the Soviet Union. Yet she spoke and read Russian, and she cooperated extensively with leading Soviet archaeologists – at the very height of the Cold War.

However, it is clear that Gimbutas's experiences as a young refugee coloured her views on how the Indo-European languages arrived in western Europe. In her conceptual world, they came with hordes of warlike men who swarmed in from the steppes to the east 5,000 years ago.

Reality was probably somewhat more complex. But despite her emotional overinterpretations, Marija Gimbutas was a pioneer, and the most recent research confirms her ideas on a number of issues.

CHAPTER TWENTY-SEVEN

DNA Sequences Provide Links with the East

W hen researchers such as Marija Gimbutas, David Anthony and James Mallory formulated their theories about the arrival of the early Indo-Europeans in Europe, they based their writings solely on archaeological excavations and comparative linguistics (and, in Gimbutas's case, traumatic experiences in her youth). They had no DNA findings to go on. But now DNA results are starting to roll in.

When I meet James Mallory in Belfast in January 2015, I have already read up on all the findings of the last few years from ancient mitochondrial DNA and the few analyses that exist of ancient Y chromosomes. They provide some support for the steppe theory, though it is not entirely convincing.

The most significant study is lying in a plastic folder on the coffee table in front of Mallory. It has been sent to *Nature*, but at the time of my visit to Belfast it is still unpublished. To my great frustration, I am not allowed to read it. For the last few months I have been hearing that this study is under way, and I have tried to persuade three of the researchers to let me see it in advance. All of them turned down my request.

Mallory refuses too. But he does tell me one thing. Since he received the study three weeks ago, it has vastly strengthened his conviction that the Indo-European languages spread to central Europe from the steppes around the Volga and the Don. And he has been working on the issue for over 40 years.

What was generally known at the time of my meeting with Mallory in January 2015 was that there were quite a lot of results from ancient mitochondrial DNA, a few results from ancient Y chromosomes, and a fairly rough genealogical tree based on the Y chromosomes of men living now.

As I explained earlier, mitochondrial DNA is a small quantity of DNA that everyone inherits from their mother.

We have seen how Guido Brandt, the doctoral student from Mainz, conducted extensive analyses of hundreds of hunters and farmers living in Saxony-Anhalt, Germany, over a period of several millennia. The analyses show beyond a shadow of a doubt that farmers from the Balkans and Hungary reached Germany in a discrete wave of immigration some 7,500 years ago. But they also point to subsequent waves of immigration. One that arrived about 4,800 years ago shows up particularly clearly in the DNA material. It coincides with a new culture in the region, known as the Corded Ware culture, which is associated with new types of pottery, flint tools and burial customs.

***

Y chromosomes occur only in men, who inherit them from their fathers. DNA from Y chromosomes is harder to extract than the mitochondrial variety. Only very few such analyses of ancient DNA have been carried out so far. The R1 haplogroup is a particular useful source of information on the wave of immigration that appears to have come from the steppes in the east.

Haplogroup R1a and its twin, R1b, are the most common groups everywhere in today's Europe. Over half the men in Europe belong to one or the other. Curiously, however, archaeological material from western and central Europe shows no sign of them before about 4,800 years ago, to judge from the DNA analyses so far conducted.

In central Europe, the oldest evidence of Y chromosomes with R1a has been found in association with the Corded Ware culture. The first analyses were published in a German study, for which Guido Brandt again did much of the work. Brandt and other researchers examined DNA from four graves in Eulau, on the River Saale. A total of 13 individuals were buried there, divided up into what look like families.

Three of the graves are exhibited at the Museum of Prehistory in Halle. I take the opportunity to look at them on my visit. The brown skeletons, with their grave goods, are displayed on a wall in the same positions they were found in. One of the labels reads 'The world's oldest nuclear family' – the title of the press release when the study on these graves was presented. A mother, a father and two sons lie close together. Both their DNA and their positions in the grave indicate that they are related. Both the man's arms are broken, and one of his sons has been struck on the head. Another grave contains what are apparently two siblings lying with their back to their stepmother or aunt. The adult woman is holding a baby to her, probably her biological child. She has been killed by blows from an axe. A third grave holds a mother and her little daughter. Another one, which is not on display, held a father and his two children.

All these people seem to have been killed at the same time. One individual has a stone arrowhead in her back, while two have fractured skulls. Several individuals have injuries to the arms, as if they had been trying to defend themselves against their enemies.

They died 4,600 years ago. Both the manner in which they were buried and the contents of their graves show clearly that the dead belonged to the Corded Ware culture. Their bodies are arranged in a typical way: the men lie on their right side, with their heads towards the west, while the women are placed on their left side, with their heads pointing eastward. Both sexes face towards the south. The children are facing their parents, and all the members of a family are placed close to each other.

Most of the dead from Eulau were women and children. There are only a few men, and they seem to have been slightly older. Researchers interpret this to mean that all the adolescents and the men in their twenties were away from home when the attack came. On returning home, they met a terrible sight; dead bodies were strewn everywhere. So the young people buried their family members with care, in accordance with the traditions of the time. They held a funeral meal that included meat. Animal bones bearing scrape marks have been found in all the graves. Beer or mead may also have been part of the tradition. The terracotta shards found may be the remains of drinking vessels. The men and boys were buried with the stone axes typical of their culture, while the grave goods for women and girls were flint tools and jewellery made of animals' teeth.

The researchers behind the DNA study note that the people of Eulau were massacred at a time of turbulence and violence in central Germany. At exactly that time, the Corded Ware culture was beginning to become common in the region, and the graves of the men and boys show a new patrilineal ancestry.

Peter Underhill, an American geneticist at Stanford University, has devoted a great deal of effort to constructing the family trees of living men belonging to various subgroups of R1a. His most recent article describes research involving over 16,000 men from 126 different populations in Europe and Asia. He has used this data to construct a genealogical tree for R1a. However, Underhill's genealogical tree is low resolution. He has not had access to any of the latest input from individuals researching their family history, he tells me by email.

This is where my uncle Anders and I come in, making our own modest contribution to science.

***

In the nineteenth century it was common for private individuals to contribute to new scientific discoveries. Charles Darwin developed the theory of natural selection at home, in his country house south of London. Mary Anning made her living from selling fossils in Dorset; though she had hardly any schooling, she discovered many pieces of evidence, which helped lay the foundations for the theory of evolution. Beatrix Potter ran a farm and wrote picture books for children, but she also presented groundbreaking new findings about fungi and their propagation.

Today's research is more technically sophisticated. It has become rare for amateurs to be able to contribute anything.

But it does happen. I have come across several examples in my time as a science journalist. Amateur astronomers discover celestial phenomena, which their professional counterparts can investigate further. The best-informed experts on particular species of flowers and subgroups of insects are often private individuals. Amateur geologists discover caves, fossils and minerals, thereby helping professionals in the field.

And now, thanks to DNA technology, people looking into their family background can help complete the picture of our common history.

I bought a new test for family history researchers called Big Y, which is sold by the company Family Tree DNA. This test was – and, at the time of writing, still is – one of the most detailed analyses of Y chromosomes that can be ordered by a private individual.

It is my own patrilineal ancestry I want to research – the line of my father, Göran, and my paternal grandfather, Eric. Being female, I have no Y chromosomes. My grandfather and my father are both dead. However, my uncle Anders agreed to have the test done. Anders is a retired history lecturer who taught in upper secondary schools and universities for many years. He and his wife had no children of their own, but when they were students in Uppsala and my four cousins were orphaned, they moved down to Lund to lend a hand. Sadly, I had relatively little contact with my uncle as a child, so I am all the happier that we have a joint project now.

My uncle and I are pioneers; we are among the hundred or so family history researchers in Sweden to take the first tests with Big Y. Our data can be combined with that of many others doing simpler tests.

Together, we contribute far more data than that available to Peter Underhill – a professional researcher – from his European material. Although he has been able to use samples from a total of several thousand men, most of his analyses are fairly low resolution. Only eight samples in Underhill's study were analysed at a really high resolution, and none of those are from Scandinavia. This means that high-resolution results from family history researchers like us can enable whole new genealogical trees to be constructed for Europe, and, in particular, Scandinavia. The trunk and most of the larger branches are based on the work of professional researchers. But our DNA helps fill in the picture. With our help, previously unknown 'side shoots' appear. Branches that were previously indistinct now take on far more definite contours.

The genealogical tree to which Uncle Anders and I have contributed shows how a man belonging to haplogroup R1a made his way towards Sweden about 4,500 years ago. He bore a new and unique mutation called Z284. That being rather dull and technical, I shall call him 'Ragnar' for simplicity's sake.

Ragnar's forefathers seem to have come from areas of southern Poland about 4,800 years ago. He himself was to be the forefather of the vast majority of men of Scandinavian descent who belong to haplogroup R1a – about one in every six men in Sweden.

The person who has contributed most towards constructing Ragnar's genealogical tree is a genealogy researcher from Härnösand called Peter Sjölund. Trained as an environment and health inspector, his day job is as a business development manager in environmental chemistry. In his free time – in the evenings, at night and at weekends – he writes about aspects of Sweden's history, using DNA analyses from private individuals.

The first time we meet, we are both speaking at a seminar on DNA for family history researchers. Peter Sjölund's talk is about how DNA analyses enabled him to dispel some ancient myths about the legendary Bure family. He shows how a great many Swedes who believe themselves to be the descendants of a particular mediaeval ancestor are no such thing. After our respective presentations, I invite him home for a pizza. I am impressed by his expertise, his capacity to absorb the new technology, and his ability to examine sources with a critical eye – just like a professional researcher.

Sjölund's preliminary conclusion is that 'Ragnar' lived in Denmark, probably in Jutland or the region that is now German Schleswig-Holstein. There are indications that my forebear Ragnar belonged to the Corded Ware culture. Feeling that I need to know more about this culture, I travel to Cracow in southern Poland to interview the Polish archaeologist Sławomir Kadrow.

***

Cracow's old town centre is one of the most picturesque you could hope to find. It is girded by the River Vistula, one of Europe's most important transport arteries for thousands of years. In the heart of the Old Town lies its famous square. It is vast, with a building in the middle where long lines of stallholders are selling amber jewellery. The cafés and restaurants around the square sell mead, the traditional honey wine that is still popular in Poland. It is autumn when I visit Cracow, and the weather is cold, so the mead is served hot in little cups. In summer it is drunk cold. It tastes very sweet, with a tinge of acidity, more or less like a dessert wine, though without the pungency lent by grapes.

I use the day before the interview I have arranged with Sławomir Kadrow to visit Cracow's archaeology museum. The first thing to meet the eye in the entrance is the world's oldest pictorial evidence of a wheeled wagon. The image adorns a terracotta vessel about 5,600 years old, which was found in the village of Bronocice, a few dozen kilometres from Cracow. What visitors get to see is a reconstruction. The real vase is in fragments, which are kept in the museum's storeroom. But the stylised image is unmistakable; the wagon has four wheels, two axles and a pole with a yoke for draught animals. The person who drew this picture must have seen wagons rolling past in real life – there's no doubt about that.

On the morning when I am to meet Kadrow, it is pouring down and most tourists are indoors. Kadrow's place of work is the Cracow section of the Polish Academy of Sciences, a large stone building just a few hundred metres away from the main square. I have to wait nearly half an hour, as the downpour has brought traffic to a standstill. But when Kadrow arrives, he takes the interview very seriously and spends the whole morning with me.

'What exactly was the Corded Ware culture?' This is my first and most important question. The need to answer it is my reason for travelling all the way to Cracow. Poland and the Vistula were the heartlands of the Corded Ware people; that is why it is so important to talk to Polish experts.

Kadrow does not hesitate for a second. 'The Corded Ware culture was a fusion,' he says. A fusion. A mix. A blend of people and cultures; that seems to be a recurrent theme. The word 'fusion' recurs constantly in the course of my interviews with the world's leading archaeologists. The most significant stages of Europe's development seem to have followed in the wake of migrations, encounters and fusions. Modern humans interbred with Neanderthals in the Middle East, the Ice Age culture known as the Solutrean developed completely new types of sophisticated tools when people were forced to live in close proximity during the most extreme period of cold, immigrant farmers merged together with groups of local hunting folk in Scandinavia, and so on.

My next question is a predictable one: 'So which groups merged to form the Corded Ware culture?'

The reply is unexpected. 'The most important, by a very long way, was the Funnel Beaker culture.'

The Funnel Beaker people! They were the farmers who were the first to arrive in Denmark and Sweden. Could the key influences have come from our direction – from Scandinavia? Some German researchers – both the DNA researchers from Mainz and Bernd Zich, the more traditional archaeologist from Halle – have been thinking along these lines. They think some of the Funnel Beaker people from the north migrated towards central Europe at about the same time as the Gökhem farmers built their passage graves in Västergötland some 5,500 years ago. Moreover, the pottery vessel from Bronocice with the world's first picture of a wheeled vehicle was found among remains from the Funnel Beaker culture.

Kadrow points out on a map the encounters between cultures and the migrations that he thinks were most significant in the development of the Corded Ware culture. One factor was the Funnel Beaker farmers from Jutland. There was a culture known as the Globular Amphora culture in south-eastern Poland, a region also influenced by the Tripolye culture from Ukraine. The groups from Jutland and south-eastern Poland met at some point or other, and resulted in the genesis of the Corded Ware culture.

And there were indeed influences from the steppes in the east as well, says Kadrow. All the evidence points to an expansion of people from the steppes, moving from east to west. From Hungary, they spread northwards to Germany and Poland. Their burial sites show a new culture and people who were 10 centimetres (four inches) taller than those already living in the region.

Kadrow's predecessor at the archaeological institution in Cracow viewed the Corded Ware culture as being simply a straightforward extension of the existing farming culture. 'But now the Corded Ware culture is regarded as something radically new. Like a revolution,' he says.

Naturally, Kadrow is familiar with David Anthony's theory that equestrian herders from the steppe introduced the forerunner of the Indo-European languages to many different areas. He is somewhat sceptical about this. In particular, he has doubts about Anthony's strong emphasis on the role of horses. They certainly became very important at a slightly later stage in the history of Europe and Asia. However, there is no evidence of horse-drawn wagons earlier than 4,000 years ago. And the earliest examples were found at a remote location in the steppe, near the River Tobol east of the Urals. Horses may well have been domesticated as early as 7,000 years ago. However, according to Kadrow they served only as a source of meat initially, not as a means of transport. Oxen and ox-drawn wagons played a far more significant role, at least for the first 1,500 years after the invention of the wheel.

Kadrow wants to emphasise another factor that he thinks was even more significant than horses when the Indo-European languages and the new ideas began to spread out over the world. It is one he thinks David Anthony and other proponents of the steppe theories have underestimated. He wants to focus on boats.

I ponder Kadrow's words for a whole year. The more facts I gather, the more convinced I am that he is right. Boats played a more important part in Europe's early history than is generally recognised. Most researchers have underestimated the role of sea voyages. If we place more emphasis on boats, several apparently irreconcilable pieces of the puzzle fall into place.

One example is the old discussion about whether the cradle of the Indo-European languages was the steppe north of the Black Sea, or Anatolia, which lies to the south of it. Both camps may be right. One of the groups involved may have been a group of farmers who started out from Turkey, travelled by boat to the northern shore of the Black Sea and settled along the Don. And it was only then that the great migration began, which took them in all directions.

It might seem a tricky thing to carry lambs, goats and calves long distances in a small boat. But the farmers who colonised Cyprus did just that, paddling more than 80 kilometres (50 miles) across open water – over 10,000 years ago.

If one accepts that people crossed the Black Sea by boat early on – and the Caspian Sea as well, no doubt – it is easier to understand that the oldest pottery in eastern Europe was found on the shores of the Don, that it is nearly 9,000 years old, that it was found in the settlements of people who kept livestock, and that it closely resembles pottery of the same era from the Middle East, south of the Black Sea.

***

Finally, the long-awaited DNA study comes out. It is published in *Nature*, and, to my relief, I manage to incorporate the results before this book goes to print. These were the results that convinced James Mallory from Belfast that he had been on the right track for the last 40 years as regards the origin of the Indo-European languages.

A large international team of researchers has compared nuclear DNA from a total of 94 individuals from Russia and the rest of Europe. The oldest of them lived in the Old Stone Age or Palaeolithic, the most recent in the Bronze Age. The researchers analysed the DNA of 69 of these individuals themselves, while the other 25 DNA sets had already been published previously.

On seeing the results, I understand Mallory's relief. The new comparison shows very clearly that the Corded Ware culture, which turned up suddenly in central Europe 4,800 years ago, was associated with a major migration from the eastern steppes – just as Gimbutas, Anthony, Mallory and many others had claimed. Three quarters of the German Corded Ware people's DNA matches that of the Yamnaya people from the pastoral culture of the Russian steppes – 2,600 kilometres (1,600 miles) away.

The Yamnaya people, in their turn, were also a mixture, bearing DNA from two separate sources. Their origins can be traced in part back to earlier Stone Age hunters in Russia. They have a distinctive north–eastern genetic signature, which researchers have found in Stone Age hunters all the way from Karelia to Lake Baikal in Siberia. However, the Yamnaya also bear genetic material from the south, from farmers who had started their journey in eastern Anatolia or Armenia. This result also ties in neatly with David Anthony's hypothesis that horses were domesticated when newly arrived farmers from the south and hunters already living in the steppe met and inspired each other.

What also emerges clearly from the new study in *Nature* is that Y chromosomes belonging to haplogroups R1a and R1b, which hardly ever occurred in western Europe before the Corded Ware culture, were very common in Russia. The type of Y chromosomes that now predominate in European men thus originated in the eastern steppes.

I recognise most of the names in the study's long list of authors. I have interviewed many of them for my research over the last few years: the first author, Wolfgang Haak, from Australia; the Swede Fredrik Hallgren, who contributed DNA samples from the Stone Age people of Motala; Johannes Krause from Tübingen; the Hungarian archaeologists Eszter Bánffy and Anna Szécsényi-Nagy, whom I met at the seminar in Pilsen; and, of course, Guido Brandt, the doctoral student from Mainz whom a more high-tech rival on the tram in Pilsen called an 'idiot' on account of the comparatively simple DNA technology he uses.

However, I have not met the main author of the study, the geneticist David Reich from Boston, or his collaborator, the mathematician Nick Patterson. The latter has a background in military code-breaking in the UK and the US during the Cold War and has also worked in finance, developing data models to predict stock exchange movements.

That combination of relatively simple DNA technology and mathematics is one of the keys to the groundbreaking *Nature* study of European history. The research team analysed a limited but carefully selected amount of nuclear DNA. Restricting their study to a comparatively small quantity of DNA freed up enough financial and other resources for them to analyse more than twice as many Stone Age and Bronze Age individuals than all previous studies put together. And it was this balance that enabled them to draw their conclusions.

Science is not just about using the most cutting-edge technology available. It is also about asking the right questions, and using exactly the right level of technology to answer them.

Only three months later another major study is published in *Nature*, led by researchers in Gothenburg and Copenhagen. They have examined even more individuals from the period in question, a total of about 100 from both Europe and Asia. The results support the theory that the Indo-European languages were disseminated by Yamnaya pastoralists from the Russian steppes. First they spread Indo-European languages to the west, throughout Europe, via the Corded Ware and Bell Beaker cultures. About 1,000 years later, a wave spread from the steppe eastwards and southwards, towards India, Afghanistan and Iran. Horses and horse-drawn wagons played a decisive role in the latter wave of migration.

The new DNA results show that the Yamnaya herders were genetically predisposed to be tall. Archaeologists had already noted their unusual height, judging on the basis of the skeletons found in graves from that period. Their height was not just a result of the large herds they kept as steppe pastoralists, giving them access to ample amounts of nutritious food; it was also a specific genetic trait.

The new DNA studies provide no support for the old notion that 'Aryans', the first speakers of Indo-European languages, were particularly blond or blue-eyed. The Yamnaya herders seem to have been quite dark, both in comparison with today's Europeans and with their contemporaries, the farmers of western Europe. On average, they seem to have been even darker-skinned than the Stone Age hunters of Scandinavia. Nearly all Yamnaya pastoralists seem to have been brown-eyed, while blue eyes appear to have been common in western and northern Europe for several millennia.

The two new *Nature* studies now provide far clearer confirmation of the picture we have sketched. Europe's population was largely formed by three major waves of immigration. The Ice Age hunters were the first to arrive, followed by the farmers from the Middle East just over 8,000 years ago. Later, about 4,800 years ago, came a wave of pastoralists from the eastern steppes. But when did this third wave arrive in Sweden?

CHAPTER TWENTY-EIGHT

Battleaxes

The local Swedish variant of the Corded Ware culture is now known as the 'Battleaxe culture', an older name being the 'Boat Axe culture'. It came into being about 4,800 years ago, and its main distinguishing feature is its characteristic stone axes. The several thousand that have been found resemble upside-down boats, hence the older name.

In the past, it was generally assumed that these axes were used to crush enemies' skulls. Old-school historians imagined hordes of rampaging men thundering into the country on horseback, battleaxes at the ready. It is an image still conveyed today by some popular history books. However, when later archaeologists examined the facts more precisely, they found no support for this view.

For four decades, beginning in the 1960s, the archaeologist Mats Malmer published a series of extensive compilations of finds from the Battleaxe period in Sweden and Norway. These provide detailed lists of the many stone axes, terracotta pots, tools, items of jewellery and other artefacts found in hundreds of graves.

One of the points he makes is that these battleaxes were not particularly useful in a battle, being so delicate that they would have cracked if anyone had tried to use them to fight. When it came to crushing someone's skull, a good old cudgel was a far superior weapon. Nor is there any evidence that the Battleaxe period in Sweden was any more violent than any other period. The skeletons found in burials from this period all belong to people who seem to have died a natural death.

Malmer concluded that the Battleaxe culture could best be compared to a religious revival. In his view, the new ideas spread like wildfire in association with the arrival of copper axes. Admittedly, no copper axes have been found in

Battleaxe culture burials, just odd items of copper jewellery. But Malmer remained convinced that copper axes existed, and that they were so prized that they revolutionised trade and culture within a short space of time. In the past, property had been collective, and it had been bound up with the whole group and the earth. Copper axes suddenly made it individual and mobile, according to Malmer's theories. He thus stressed the significance of trade and religion. In his view, immigration had no significant impact on the course of events.

Some of today's archaeologists agree with Malmer that the Battleaxe culture was barely anything more than a gradual development among people already living in Sweden and Norway. 'The ceramics were new. But in other respects one can hardly speak of significant inward migration; it was more a continuation of an existing culture,' says Karl-Göran Sjögren from Gothenburg, for example.

Others give more credence to outside influences. One of them is Lars Larsson, a professor emeritus from Lund, who, in other contexts, prefers to downplay the role of immigration in early Swedish history. The Battleaxe culture, in his view, is an exception. Larsson believes that its exponents lived mainly as pastoralists and that they were far more mobile than the earlier farming population. Their nomadic lifestyle, he argues, explains why archaeologists have found comparatively few settlements reflecting this culture. Even small groups of immigrants may have had a major impact on development, he thinks, introducing a new economy and ideology.

Åsa Larsson from Uppsala, who wrote her doctoral thesis in 2009, turns these ideas upside down. She also thinks the Battleaxe culture was spread to some extent by immigrants. However, in her view they came not from the lands south of Skåne, but from Finland and the Baltic. And they were not men on horseback with battleaxes at the ready, but women with specialised skills in pottery. They came by boat via the Åland archipelago to marry men living in Uppland and Södermanland. From these provinces, the Battleaxe culture subsequently spread to southern and western Sweden and to Norway, according to Åsa Larsson's theory.

Her conclusions are based mainly on the appearance of the ceramics and the burials of this culture. Within the Corded Ware culture, there are considerable similarities between Ukraine and the Netherlands. But there are also some regional differences. Corded Ware people everywhere had strict rules on how the bodies of men, women and children were to be positioned in their graves: on which side they were to be laid, in which direction the head was to face, and where battleaxes and terracotta vessels were to be placed. However, these strict rules varied somewhat from place to place. Similarly, there were subtle variations in their pottery.

Corded Ware pottery differed in several respects from that of earlier cultures. The pattern, embossed with twined cords, was new, and it was applied in a very uniform fashion. The most common type of vessels, by a long way, were small hemispherical cups.

Another typical feature was that the Corded Ware people mixed clay with crushed fragments of pottery from older pots. Mixing clay with hard material is known as adulteration, and potters have used different types of material for this purpose almost since the birth of ceramics: sand, grit, crushed seashells, chaff and so on. Incorporating crushed pottery, however, is a typical feature of the Corded Ware/Battleaxe culture.

Researchers in Finland with whom Åsa Larsson collaborates have shown that Finnish pots from the Battleaxe period sometimes contained crushed pottery from Sweden. And the same applies in reverse; there are Swedish ceramics incorporating fragments of pottery from Finland. Such pottery, made of clay from Sweden mixed with pottery fragments from Finland, has been found as far south as the area around Kristianstad. Åsa Larsson believes the Corded Ware people saw a deeper meaning in mixing crushed fragments of old pots into new ones. This was a way of making links with older and dead kinsmen and of imbuing ceramic vessels with eternal life.

To be nitpicking, Åsa Larsson has relatively little support for her view that it was women who produced the terracotta vessels of the Battleaxe culture and migrated over the Baltic Sea. She, however, thinks this was the case, partly because women are the main potters in contemporary cultures that anthropologists have studied.

A number of researchers, including Marija Gimbutas, have claimed that the early Indo-European cultures of the steppes were strongly male-dominated societies. Finds from the eastern regions of the steppe support this view; the finest graves belong almost exclusively to men. By contrast, the burials of the Swedish Battleaxe culture show no difference in status between the sexes. Both were given grave goods in the form of terracotta vessels, bone and stone tools, and haunches of roast venison and mutton.

The big difference seems to be that only men and boys were buried with battleaxes. Men were accompanied by large, beautifully polished axes and small boys by miniature versions. Women, on the other hand, seem to have been given jewellery such as amber rings and beads, and, in a few cases, copper bracelets and beads.

***

Kristian Kristiansen is Danish but works as a professor of archaeology in Gothenburg. At least four other archaeologists warned me about him. 'He has a very broad-brush approach,' is the most diplomatic comment. 'He has his theses which he wants to try and prove,' says one. 'He was good when he was younger, but now he's lost himself in Vedic scripts,' claims another.

I read a series of articles and a substantial book by Kristiansen before going to Gothenburg to interview him. The comment on his broad-brush approach turns out to be true; he seeks to discern patterns across millennia and extensive geographical areas. In other respects, however, I think his peers' judgements are unfair. Kristiansen is exceptionally well read. I find him less dogmatic than many other archaeologists I have met as regards his views on both migrations through history and DNA. And I like the fact that he does not speak ill of his competitors – unlike many of them. The most critical remark he makes is that some of his peers 'have a very local perspective'.

Like Lars Larsson, Kristiansen thinks the Battleaxe culture was associated to some extent with immigration, but that the immigration was limited in scale. The immigrants nonetheless became very important through their networks with similar Corded Ware cultures elsewhere in Europe. The people who arrived with the battleaxes brought with them new deities, new rites, new tales and songs, and a new relationship with property, inheritance and rights – in short, an entirely new social structure.

The people of the Battleaxe culture came to Sweden by water, and they may have arrived from different directions. Åsa Larsson may be right; some may have come from Finland and the Baltic countries. But others came from the south, from Poland, Germany and Denmark. Kristiansen believes that an Indo-European language arrived in Sweden at around that time and that it was probably an early form of Gothic. Mastery of that language was part and parcel of belonging to the new network.

Personally, I would like to leave open the question of whether this language was Gothic. After all, we are talking about a period almost 5,000 years ago, and the oldest samples of Gothic are only about 2,000 years old.

The pattern was far clearer in Jutland, says Kristiansen. Pollen analyses conducted there show a dramatic change in the landscape that started about 4,800 years ago. In the course of only a century or so, large quantities of trees disappeared and were replaced by open grassland – exactly what you would expect when an expansive pastoralist culture becomes prevalent.

The burials typical of this culture emerged during the same period. The Danes call their local variant of the Battleaxe culture the 'Single Grave culture'. The name is based on the obvious difference between burials belonging to this culture and those of the earlier farming culture, in which the same burial chamber was used for a group of people. Graves belonging to the Corded Ware/Single Grave culture mostly held just one individual or at most a very few people. Researchers generally interpret this different burial pattern as signifying that individuals and their immediate family were starting to play a more important role, while the group and collective interests were declining in importance.

Although Kristiansen is nearing pensionable age, he is definitely not one of the band of older archaeologists who regard DNA technology with suspicion. In fact, he has applied for and been awarded a huge EU grant in cooperation with leading DNA researchers from Copenhagen.

These researchers have analysed over 100 individuals from Europe and Asia. In June 2015 they published their results in *Nature*. The Danish–Swedish team confirms the picture drawn by their competitors in Boston a few months previously: about 4,800 years ago, a substantial wave of immigration reached central Europe from the Russian steppes. With it came a whole new culture and, almost certainly, Indo-European languages. Pastoralists from the Yamnaya culture of the steppes migrated both westward and eastward, even reaching China. This was followed somewhat later by southward migrations that dispersed Indo-European languages to India, Afghanistan and Iran.

The Boston team and their competitors from Gothenburg and Copenhagen have reached exactly the same conclusion: European prehistory can be summarised in terms of three substantial waves of immigration.

The Danish–Swedish team has also been able to provide information about the Battleaxe culture, the subject of disputes between archaeologists for many decades. The 100-odd individuals whose DNA they analysed include several from Sweden, one of whom was a man buried about 4,500 years ago in a Battleaxe culture grave in Viby, near Kristianstad.

This man has genetic traits resembling those of people from the Yamnaya culture. His DNA shows he originated from the steppes of eastern Europe, just like the Corded Ware people of Poland and Germany. His Y chromosomes belong to haplogroup R1a. The material also includes a slightly more recent member of the Battleaxe culture from Lilla Beddinge near Trelleborg. He, too, shows strong Yamnaya genetic traits and belongs to haplogroup R1b.

This means the legendary archaeologist Mats Malmer was mistaken. DNA researchers have finally settled the old controversy that divided Swedish archaeologists for so long.

It is now clear that immigrants brought the Battleaxe culture to Sweden. There may not have been large numbers of incomers initially, but they played a decisive role in the fundamental social changes that began to take place. With their common origin, culture and language, they were able to maintain networks extending over vast distances.

It is unclear whether the first Battleaxe people in Sweden actually brought horses with them. Only a few horse bones have been found, and it has not been established that they belonged to domesticated horses. The knowledge we have today means the old image of hordes of violent men riding into Sweden with their battleaxes at the ready needs to be taken with a large pinch of salt.

Yet the arrival of these people really did herald the start of a new age and a completely new way of life. That lifestyle was based on large herds of livestock and ox-drawn wagons in which herders could spend the night for long periods. The package also included metal, woollen fabrics and mead, and powerful men with numerous offspring. More than half of the men in today's Europe – perhaps closer to two thirds – are the direct descendants of such clan chieftains.

'Ragnar', the man from the Battleaxe culture who was my forefather and the forebear of every sixth Swedish man, probably still lived outside Sweden 4,500 years ago. He is most likely to have lived in Jutland, as part of the expansive pastoralist culture that had taken over there. This is shown by the genealogical tree of haplogroup R1a, which my uncle, Anders, and other amateur family history researchers have contributed.

It is likely that several centuries passed before Ragnar's descendants made their way over to Sweden. However, once they had taken that step, they changed society at least as radically as the first farmers had done a few millennia earlier.

CHAPTER TWENTY-NINE

Bell Beakers, Celts and Stonehenge

A few centuries after the Corded Ware people had dispersed throughout eastern Europe, a similar but slightly different culture spread across the western part of the continent. Known as the Bell Beaker culture, it paved the way for the Bronze Age.

The British Isles were a particular stronghold of this culture. The Bell Beaker people are believed to have played an important role in the legendary monument of Stonehenge. A few kilometres from the stone circle, archaeologists have found graves containing the remains of a number of Bell Beaker individuals.

One of these is known as the Amesbury Archer. He was about 40 years old at the time of his death about 4,300 years ago. His grave is regarded as the most valuable find ever from the British Bronze Age. The gifts that accompanied him in death were very costly, comprising four ceramic vessels, three copper knives of different sizes and two hair ornaments made of gold (the oldest gold artefacts ever to be found in Britain). A slate wristguard associated with archery was found next to his arm, along with 16 arrowheads. A similar one lay at his feet. His hide cloak was held together by a large bone needle. A number of small bone and flint tools had been placed in a pouch, although this has rotted away.

However, the most interesting tool of all is a small stone anvil. British researchers interpret the anvil, the copper knives and the gold ornaments as showing that the man was a craftsman who specialised in metalwork. They believe the Archer's expertise in this field would have given him an elevated social status.

A nearby grave contained a man in his twenties who appeared to have been the Archer's son or younger brother.

Both of them had a rare hereditary anomaly of the heel bone. The younger man had similar hair ornaments made of gold. He was clearly born locally; that can be seen from the composition of the isotopes in his teeth. However, the older man, the Archer, turned out to have been born somewhere quite different and distant from Stonehenge. To the researchers' amazement, he appears to have spent his childhood in the Alps or Germany.

Clearly, some Bell Beaker people were very mobile, travelling not just along the western seaboard of Europe, but also along watercourses and other routes that extended well into central Europe.

***

Just as Y chromosomes belonging to haplogroup R1a are very common in eastern Europe, where the Corded Ware culture was at its strongest, the twin haplogroup R1b is best represented in the areas where archaeologists have found most traces of the Bell Beaker culture.

The US genetics professor Peter Underhill has examined thousands of samples from men living today and constructed genealogical trees for R1a, as I mentioned on page 257. He has also constructed similar genealogies for haplogroup R1b. In Underhill's view, it is quite conceivable that a major subgroup of R1b, known as M412, was dispersed in association with the Bell Beaker culture. At any rate, the time of its dispersal fits perfectly.

The latest analyses of DNA from archaeological remains and family history researchers now confirm this hypothesis. Just as with R1a, lineages that include R1b appear to originate with the Yamnaya culture from the Russian steppes. It was thus from the steppes that men with the R1b haplogroup spread out over western Europe. The most reasonable explanation is that they travelled by boat.

Today, R1b (with its subgroup M412) is the most common group of Y chromosomes in the whole of Europe, followed by R1a. More than half the men in Europe belong to one of these two groups. In Sweden, more than one in three men belongs to either R1a or R1b.

Many researchers believe that the Bell Beaker people brought with them an early variant of the Celtic languages. Today, this group of languages survives only in the far west of Europe, being confined essentially to Brittany (France), Wales, Scotland and Ireland. However, before the time of the Roman Empire, the Celtic tongues were far more widespread and were spoken through much of western Europe.

***

The British press came up with their own name for the man in the lavish grave. They dubbed him 'King of Stonehenge'. Of course, we cannot know exactly what links he had with Stonehenge. But it is clear that his son or younger brother was born nearby, and that his own lifetime coincides with the period when some of the great stones were raised.

This huge megalithic construction is now Europe's best-known prehistoric monument. I visit it one icy January morning. This time I am a good deal luckier than on the cloudy day when I visited the Goseck solar observatory in Germany. At Stonehenge, the dawn sky is limpid. I am with a small group of tourists on one of the few special tours that enable you to enter the stone circle. Like all visitors, we come by bus from the entrance to the visitors' centre. A few minutes before sunrise we leave the bus and walk a few hundred metres towards the great stones. By the time the first rays of the sun glint on the horizon, I am inside the circle. The light is framed by two huge stones – a stirring sight.

Annoyingly, I am absolutely freezing. The cold January winds sweep over the open landscape. Wrapping my woollen scarf around my head a few times, I console myself with the thought that this wintry weather is part of the experience. These days, Stonehenge's high season is midsummer, when entry is free and thousands of tourists – including neo-pagans and self-appointed druids – flock around the stones to mark the summer solstice. But sunrise at the summer solstice lies on the same line as sunset at the winter solstice. Today, leading British archaeologists think the megaliths were raised primarily to celebrate the winter solstice – just as with the Goseck solar observatory, built 1,500 years earlier. Within these structures, the people of the time celebrated their version of Christmas or Yule.

The ambitious guided tour continues throughout the morning. We walk around the site and are given a detailed account of the ways in which Stonehenge was used over the millennia.

In recent years, archaeologists have mapped a large area around the monument. Vincent Gaffney, the lively professor I tried to interview in a hotel bar in Bradford, has led an extensive research project using a range of remote analysis tools. The collected findings show the area was first used by Stone Age hunters about 10,000 years ago. At the site now occupied by the car park, researchers have found traces of three pits that once contained long pinewood poles. These were lined up with a tall tree. According to one interpretation, they may have been totem poles of some kind, with ritual significance.

About 5,500 years ago, when agriculture had just reached England, people began to build structures including several long, narrow burial mounds, a massive elliptical ridge and several round formations comprising ditches, earthworks and wooden posts. The circular formations are similar to those at Goseck and other solar observatories in Hungary and Germany built by early farmers. A similar wooden structure, called Woodhenge, has been reconstructed a few kilometres north of Stonehenge. As the new posts are made of concrete and are only about a metre (3¼ feet) high, the reconstruction lacks the imposing atmosphere of the Goseck observatory, with its tall oaken posts.

Near Woodhenge, archaeologists have found traces of a whole village, known as Durrington Walls. The small houses there were first used about 4,500 years ago, but only for a short period of the year – just before, during and after the winter solstice. The people who occupied the houses seem to have come on foot from all directions, bringing cows with them. They also had access to herds of pigs, which were shot with bows and arrows. Huge numbers of arrowheads have been found from the pig hunt – though wholesale slaughter would be a more accurate description – that seems to have been part of the winter solstice rituals.

British archaeologists believe these people celebrated sunrise at the winter solstice in a structure that lay near the little village and next to the river. Then they walked up in a procession towards the place where Stonehenge now lies. They may have covered part of the distance by boat along the River Avon. The route they took varied somewhat at different periods. Once the people reached Stonehenge, they would have watched the sun sink between the mighty stones. This, at any rate, is the archaeologists' latest interpretation, our guide explains.

Originally, Stonehenge would have looked more or less like Goseck, with a circular ridge on the outside, a circular ditch and an inner circle of wooden posts. It was one of several similar round timber structures in the area, according to the latest findings from remote analysis. But about 4,500 years ago people began to enhance this particular structure by raising megaliths.

One of the stones – the Heel Stone, standing about 70 metres (230 feet) outside the circle – was there from the beginning. The others were hauled there in different stages. The very largest megaliths, the sarsens, which are made of sandstone and weigh up to 40 tonnes (88,000 pounds), come from a site a few dozen kilometres away. But the smaller stones, weighing up to four tonnes (8,800 pounds), were brought all the way from Wales, a distance of about 250 kilometres (155 miles). They are made of various types of dark rock whose collective name is bluestone. These dark stones glisten in a very particular way, especially if polished and moistened.

At the time when the Bell Beaker people arrived in England and the Amesbury Archer was alive, the people at Stonehenge were thus engaged in enhancing their ancient solar observatories by adding megaliths. They shot pigs at the winter solstice, and they had begun to use metal.

This was an age of decisive importance for Europe's development. It witnessed the diffusion of the Bell Beaker culture, and metals were becoming ever more important. Some men became powerful and had large numbers of offspring; that is why the traces of these changes can still be seen in Europeans' genetic material. The Y chromosomes belonging to haplogroup R1b are one such marker.

The summer and winter solstice had been important reference points ever since farming first came to Europe. Everything suggests that early Bronze Age society attached at least as much importance to the course of the sun.

Established archaeologists in Sweden tend to be wary of engaging in astronomical interpretation of this nature. My impression is that they are afraid of being lumped together with self-proclaimed experts who have wild and highly unscientific theories about astronomy and archaeology. But perhaps they are a little too fearful. They risk throwing out the baby with the bathwater and missing important factors that researchers in other countries take very seriously.

It is not only British archaeologists who see a link between heavenly bodies and prehistoric monuments. In Germany, the most important find of the whole Bronze Age is also associated with astronomy.

CHAPTER THIRTY

The Nebra Sky Disc in Halle

The Bronze Age can be said to have begun when the Bell Beaker culture in the west and the Corded Ware culture in the east merged, producing the Unetice culture.

Initially, the Bell Beaker people and the Corded Ware people rigorously maintained their respective defining features when living in proximity to one another. Their ceramic vessels were shaped and patterned in different styles. While the Corded Ware people buried women lying on their left side and men on their right side, both facing south, the Bell Beaker people did precisely the opposite. They placed women on their right side and men on their left side, both facing north.

But about 4,300 years ago, such distinctions began to melt away, and the Unetice culture began to emerge. While this culture has left most traces in the Czech Republic and Slovakia, it extended from Germany to Ukraine and as far south as Austria.

The most remarkable find from the German Unetice period – and, indeed, the whole German Bronze Age – is the Nebra sky disc. The way this was discovered might have come straight out of a detective novel. But it's a true story.

In 1999 two treasure hunters with a metal detector went looking for spoils near the German town of Nebra. Using a metal detector without a licence is classed as a crime in Germany, as it is in Sweden, and dealing with finds in the way these men did is indisputably criminal. They discovered a bronze disc ornamented with gold, the size of a substantial pizza plate, together with two bronze swords, two bronze axes, a chisel made of the same metal and a spiral bracelet in bronze. The day after, they sold all these finds for 31,000 Deutschmark to a receiver of stolen goods in Cologne.

Over the next two years, the artefacts were sold on several times for up to a million Deutschmark. In 2001, the bronze disc was offered for sale on the black market at 700,000 Deutschmark. However, Harald Meller, the chief antiquarian in the federal state concerned, had been tipped off about the sale. He put in a bid and arranged a meeting with the sellers in the Swiss city of Basel. The police secretly monitored the transaction and arrested the sellers, and after further investigation the original treasure hunters were detained. They ended up serving prison sentences.

The Nebra sky disc is spectacular – so much so that some experts initially claimed it must be a modern forgery. It weighs several kilos. The disc itself is made of bronze covered in green verdigris as a result of oxidation. Pieces of gold plate are attached to the bronze disc. These represent a new moon, a full moon and a cluster of stars. Of the three pieces of gold plate that once adorned the edges of the disc, one is now missing.

Meller and his staff have had a series of tests conducted to check that the find is genuine. One such test involved comparing earth on the disc with soil from the place where the treasure hunters say they found it. And the results appear to confirm that it was found on the top of the Mittelberg, near the little town of Nebra, a few dozen kilometres south of Halle.

The researchers have also studied the design of the swords, the isotopes in the metal, and impurities in the gold and the bronze alloy. They discovered that the treasure was buried about 3,600 years ago, though the individual artefacts had been in circulation for centuries by that time. The copper in the bronze disc appears to have come from the Alps, not so very far from Nebra, but the gold is from Cornwall in England.

The two pieces of gold plating that covered part of the edge of the disc were not there from the outset, but were added later. They span an angle of 82 degrees, which happens to be the exact angle between sunset at the summer solstice and sunset at the winter solstice at the latitude of Nebra. The sky disc thus served to predict both the winter and the summer solstice, the turning points of the year. It had the same function as the great solar observatory at Goseck and the stone circle of Stonehenge. But the sky disc is so very much smaller – a portable instrument that could be taken out when it was required to make observations and hold ceremonies.

At an even later stage, craftsmen added an additional piece of gold plate to the margin of the sky disc. This resembles a stylised miniature boat. The gold in the stars, the two 82 degree angles and the stylised boat was processed at different times and contains different proportions of silver. It was clearly added by three craftsmen on three separate occasions.

In 2012 Harald Meller nominated the sky disc for recognition by UNESCO, and a year later it was proclaimed an item of World Heritage. The statement justifying the decision describes it as 'one of the most important archaeological finds of the twentieth century' and continues: 'It combines an extraordinary comprehension of astronomical phenomena with the religious beliefs of its period, that enable unique glimpses into the early knowledge of the heavens.'

***

I get to see the sky disc on a visit to the Museum of Prehistory in Halle. It is a very solemn occasion. As a visitor, I first have to pass through an antechamber with information about the context of the find. Then I can enter the 'holy of holies', as my companion, Bernd Zich, puts it. The room is very dim, and I practically have to fumble my way around the display case in the centre by touch. Its interior is all the brighter; the dramatic illumination gives a sense of the magic powers with which the people of the time would have endowed the gilded bronze disc. In the darkness next to the wall sits a museum attendant, watching my every move. Another, smaller case on the wall displays the other items found in the same cache: a bracelet, a chisel, two axes and two swords. Such doublets – like twins – are a common feature of Bronze Age finds.

Bernd Zich is a department head at the Museum of Prehistory. He features in Chapter 21, in which I noted that he had discovered the world's oldest known wheel tracks. Now he explains how leading German archaeologists interpret the sky disc.

When the first farmers arrived in Germany, astronomy was already very important to them. It was linked with the farming year; they wanted to keep track of phenomena such as the winter and summer solstices and the vernal and autumn equinoxes. Many large structures in Hungary and Germany, including Goseck, have been interpreted as solar observatories, which served to enable people to observe and celebrate the winter solstice. Archaeologists in England believe the earliest formations built near Stonehenge had a similar role.

But the Nebra sky disc is about 3,000 years more recent than the Goseck observatory. Human knowledge of the movements of celestial bodies developed over these three millennia. Some ingenious people managed to find a solution to an awkward problem: the fact that the solar year and the lunar year are not a perfect match. Twelve lunar months are 11 days shorter than a solar year. This lack of harmony must have appeared baffling at one time.

If you add an additional lunar phase every three years, the solar and lunar years more or less keep pace. In Zich's view, this would have been secret knowledge of great religious significance. The most powerful leaders had their own 'astronomers' who were able to interpret the course of the sun, moon and stars. These experts knew – partly thanks to the sky disc – when it was time to add an extra month.

One way of dealing with the problem is described in a cuneiform text from Babylon that dates back around 2,700 years. This states that you need to add an intercalary month every spring when a particular phase of the moon – not a new moon, but a crescent that has been waxing for a few days – is near the Pleiades star cluster.

The Nebra sky disc depicts exactly this position, with seven gold dots, the number of stars in the Pleiades that are visible with the naked eye, next to a crescent moon of exactly the right size. Those able to interpret the message of the bronze disc thus knew, about 4,000 years ago, what the Babylonians set down in writing over a millennium later.

Greek writings that are approximately contemporaneous with the Babylonian ones also mention the Pleiades as an important marker showing the time to plough and the time to gather in the harvest.

Those wishing to immerse themselves in the astronomy of early European Bronze Age cultures can make their way from the Museum for Prehistory in Halle to Nebra, which lies a few dozen kilometres to the south-west. On an elevated site very near the spot where the sky disc was found on the Mittelberg, the regional authorities have set up a large, lavish information centre. This contains a planetarium and an exhibition designed both for children and for adults with a specialist interest. One of the things I learn there is that the site where the disc was found on top of the Mittelberg was a place of worship and burial for several millennia – from the Palaeolithic to the Iron Age – before it fell into disuse.

The astronomers who mastered the art of interpreting the sky disc were probably in the service of a rich individual with a great deal of power. At the time of the Unetice culture, society was beginning to become far more stratified, and the most important leaders were starting to acquire more and more wealth, power and status. One of the signs of this is the burials known as 'chieftains' graves' (*Fürstengräber* in German), large tombs containing valuable gifts.

Dalia Pokutta, an archaeologist based in Gothenburg, has recently written a thesis on her study of the diet of a number of individuals buried in chieftains' graves. Studying isotopes has enabled her to discern whether they ate much meat and other forms of protein, or whether they lived mainly on plants. Rather surprisingly, her results show that the people buried in these large, lavish tombs had an almost ascetic diet. She compares them to Indian yogis. Her interpretation is that those interred in chieftains' tombs were regarded as priests or shamans, rather than as people whose power was based on their economic assets.

The graves Dalia Pokutta studied Dalia Pokutta are in Poland. They are part of the Unetice culture that lay behind the sky disc. Her thesis and many other signs suggest that the early Bronze Age societies of Europe were what we now call theocracies, in which religious leaders had a very strong influence. Priests played an important role. But the same was true of traders, warriors and artists.

CHAPTER THIRTY-ONE

The Rock Engravers

Not many researchers get to have their findings presented in the form of a wall-to-wall carpet. But that honour has fallen to Johan Ling, an archaeologist at the University of Gothenburg. The carpet occupies a whole room at the Vitlycke Museum in Bohuslän. Visitors walking around it can trace Europe's Bronze Age trade routes.

The carpet was inspired by an article, which Ling published in 2013. Rarely has a Swedish archaeological study attracted so much attention. Suddenly the Bronze Age appeared in a new light.

The article presented the first ever incontrovertible proof that both people and goods had circulated on a far larger scale than had previously been known. Older generations of archaeologists were only able to study the appearance of metal objects. The conclusions they drew were based on shapes, ornamentation and their own imagination. But the latest technology has made it possible to study isotopes and trace elements in metal, as German researchers have done with the Nebra sky disc. Ling is now able to show that there were links between Sweden and the Mediterranean, the Alps and the Atlantic all of 3,600 years ago.

Just take the axehead found in the River Jösse outside Arvika in the province of Värmland. Made of solid bronze, it weighs about 1.5 kilos (3⅓ pounds). There is a hole in it for the shaft and, though made of bronze, it is similar to a stone axe in shape. Researchers have somewhat different opinions on whether bronze axes of this type were used for practical tasks, or whether they were essentially status objects. Such axes may have had standardised weights, which would mean they could be used as bronze ingots representing a specific value. The axe type is known as a Fårdrup axe, after the place in Denmark where they were cast. Ling himself thinks this particular one is more likely to have been cast in Värmland, even though no hearths used for casting metal have yet been found there. At any rate, archaeologists agree that it must have been cast in Scandinavia.

Copper ore occurs in areas to the north of Arvika, and you might well imagine that the copper in an axe found in the River Jösse would come from that region. Yet that is not the case. Isotopes in the metal reveal that the copper is not from Värmland at all – but from Cyprus.

At that time there was a great deal of copper mining on Cyprus. Copper ingots from the island went on long journeys. The copper that ended up in the axe, found in the River Jösse, was probably shipped via the Bosphorus and the Black Sea, then along some of the great rivers of Europe – perhaps the Danube or maybe the Dnieper.

Another axe from more or less the same period was found on the island of Öckerö, in the northern part of the Gothenburg archipelago. It is also cast in solid bronze, though the technology used was more sophisticated. This axe is of a type known as a palstave – and we are not just talking shiny status symbols any more. Palstaves were definitely designed to split lengths of timber into planks. They were highly effective tools with a sharp, curved blade. The shape of the axe reveals that it must have been cast in England or France. The copper, however, is from neither place – it comes from Lavrion in Greece. Raw copper was probably transported from the Mediterranean up one of the French rivers, the Rhône or the Garonne. It was then unloaded in England, mixed with tin and cast to make the latest model of palstave.

Ling has so far investigated about 40 bronze objects found in Swedish provinces including Dalsland, Bohuslän, Halland, Småland and Skåne. They have been shown to contain copper from the Alps, Spain, Portugal and Sardinia, and, in some cases, even from Greece and England.

***

Bronze is an alloy of copper and tin. There is very early evidence for copper mining in Europe. Copper artefacts found in Serbia and Romania are just as old as agriculture in those regions, going back about 8,500 years. The people of the Varna culture on the shores of the Black Sea in Bulgaria could already smelt copper and gold 6,600 years ago. This is shown by a number of spectacular burial finds. One such artefact is a large penis sheath in gold, belonging to a man in one of the graves at Varna, and the grave also contains other gold objects weighing about six kilos (13 pounds) in total. Ötzi the Iceman, who died in the Alps about 5,300 years ago, was carrying an axe made of copper with a high arsenic content. Such copper becomes harder and easier to cast than other types, but it is no match for bronze axes.

It is unclear where people first learned how to purposely mix tin and copper to make bronze. A number of early bronze artefacts about 5,000 years old have been found in central Asia and the Mesopotamian city of Ur in Iraq.

To make bronze, you need tin, and tin mines were far rarer than copper mines. There were only a few places where tin could be mined. One such area was Cornwall in southern England. Several pieces of evidence have now emerged that suggest large-scale tin production began in the region around 4,200 years ago. That was just after the Amesbury Archer had died and the timber structures at Stonehenge had started to give way to megaliths. All the early European bronze artefacts analysed so far contain tin from Cornwall – including the gold plates in the Nebra sky disc.

The map on the carpet in Vitlycke Museum shows networks extending over much of Europe. They included the rivers of central Europe, just as old-time archaeologists believed. But the maritime routes along the Atlantic coast were at least as important. This can be seen from all the artefacts found in Denmark and Sweden but originating in Spain, Portugal, France and England. In view of how ocean currents flow, it seems quite clear that one of the destinations would be the west coast of Sweden. Paddling from southern England to the western Swedish seaboard essentially meant following the ocean currents.

These long-distance trading voyages called for a new type of boat. One decisive step was the technology of building boats from planks, rather than making them from hide or dugout tree trunks. The Danish National Museum in Copenhagen houses a vessel built using this method, known as the Hjortspring boat. Though it was built in the early Iron Age, around 350 BC, we have reason to believe that the people of the Bronze Age were already able to build such boats.

It is absolutely clear that the Hjortspring boat was designed for battle. Alongside it, in the bog where it was discovered, 169 spears and lances, 11 swords, a number of pieces of chain mail and the remains of about 80 shields were found. The vessel's prow and stern were identical, and it was perfectly symmetrical in structure. The oarsmen simply had to swivel 180 degrees to change direction at lightning speed, and they could disappear in the same direction they had come from.

Danish researchers have had a team of elite rowers test a replica. The athletic men managed to row nearly 100 kilometres (60 miles) in a single day. At that speed, a vessel can reach England from the coast of Jutland in under a week. The English Channel is just over 30 kilometres (20 miles) wide, a distance that a good team of rowers can cover in half a day.

Cutting lengths of timber into planks of exactly the right thickness called for axes. While stone and copper axes were some help, the big breakthrough came with the bronze axes. Once the technology of casting artefacts in bronze was well established on the Atlantic seaboard, things moved very fast.

A virtuous circle came into being: better metal axes enabled people to split timber into planks and build more seaworthy vessels. Better vessels made maritime transport faster and safer, enabling people to exploit newly discovered deposits of copper and tin. New mining sites provided raw materials for more bronze axes, which became more and more effective, and so on.

The result was an explosive development in the direction of a new type of society – one in which wealth, trade and aristocracy became far more important than in the past. Against this background, it is not surprising that certain men came to have a very large number of descendants – so many that they account for over half of Europe's modern-day population. Boats and metal can account for the dispersal of haplogroup R1b along the Atlantic seaboard and for the fact that the dispersal got under way at the time of the Bell Beaker culture.

The fact that R1a became so common further east is also linked with the successful pastoralist cultures that already existed at the end of the Stone Age, and doubtless also with the domestication of horses. But the best explanation for the fact that my forebear 'Ragnar' became the forefather of nearly one in seven family history researchers with Swedish ancestry is the Bronze Age trading network. This is the picture that emerges when, with Peter Sjölund's help, I study the dispersal of men with the R1a-Z284 mutation, descended from the individual I have dubbed 'Ragnar'.

A year has passed when Peter Sjölund and I meet for the second time. It is a sunny Sunday in October 2014, and we are at the Genealogical Association in Solna, looking at maps produced by some Russian enthusiasts. The Russians are genealogy researchers with an IT background who have set up a firm called YFull specialising in the analysis of Y-chromosomal DNA. They have collected data from a large quantity of published research, but also from private family history researchers, particularly the pioneers who have taken the 'Big Y' test.

On the maps we can see that 'Ragnar' appears to have lived somewhere near Denmark, probably in Jutland or Schleswig-Holstein. His lineage was there 4,500 years ago, at the time when the Corded Ware culture was dominant in this region. From there, one of Ragnar's descendants moved across to Sweden some 3,900 years ago. This must have been a sea voyage, and he probably landed on the west coast. It looks as if he arrived at exactly the same time as rich treasures of bronze began to make their appearance in Sweden.

The Bronze Age man who came here about 3,900 years ago would eventually have a huge number of descendants of his own, including my grandfather Eric, my father and my brother. Our forefather probably spoke an Indo-European language, which may have been a kind of Gothic, as the archaeologist Kristian Kristiansen believes.

***

When the trade in bronze reached Scandinavia, the people living there obviously had to pay for all the ingots they purchased. One of their main means of exchange was amber.

This golden-brown, translucent stone could be found on sandy beaches along the west coast of Jutland and in the southern Baltic. Amber was already seen as valuable and special in Palaeolithic and early Neolithic times. Amber beads and other objects often featured among grave gifts in Scandinavia.

But just as bronze started to become increasingly common in Scandinavia, people there stopped burying amber grave gifts. Instead, large quantities of amber have been found in graves and hoards elsewhere in Europe. And the most extensive finds of Bronze Age amber are at exactly the same locations as some of the main copper mines.

At the beginning of the Scandinavian Bronze Age, amber came mainly from western Jutland. However, Scandinavian Bronze Age rulers seem to have taken control over other amber-rich regions further east, along the shores of Poland and Germany. There also seems to have been a network of maritime routes across the Baltic. At the end of the Bronze Age, about 3,000 years ago, a bronze workshop emerged in Hallunda on Lake Mälaren, now a suburb south of Stockholm. In Bronze Age times, Hallunda was an important settlement on a sea inlet.

The carpet in Vitlycke Museum allows you to follow trade routes in all directions, with flows of metal going up to Scandinavia and amber from the Baltic flowing in the other direction. The routes from the Mediterranean, the Black Sea and the Alps ran via major rivers such as the Elbe, the Danube and the Vistula, and, above all, along the Atlantic coastline.

And Cornwall, in England, was a nerve centre in this trade. The tin mine – a rarity – was in Cornwall, and it was there that bronze ingots were produced.

Three shiny yellow substances were greatly valued during the Bronze Age: gold, bronze and amber. There must have been other important wares as well. The wrecked vessel known as the Uluburun provides unique evidence from the eastern Mediterranean. The Uluburun was found just off a cape near the Turkish town of Kaş. On the basis of the annual rings in its timbers and other evidence, it is estimated to have sunk over 3,300 years ago.

The cargo of the Uluburun included 10 tonnes (22,000 pounds) of copper and a tonne (2,200 pounds) of tin – exactly the right proportions to cast 11 tonnes (24,250 pounds) of bronze. There was amber from the Baltic, gold, and semi-precious stones such as quartz and agate. There were also over 100 terracotta jars containing goods such as glass beads, resin, olives, almonds, pine nuts, figs, grapes and pomegranates. Other luxury items included ivory, hippopotamus teeth, tortoiseshell and ostrich eggs.

Less is known about the goods that were traded from northern Europe, apart from amber. Halle an der Saale in Germany contributed salt from its salt mines. In all probability, furs and slaves were shipped southwards from Scandinavia. However, salt, furs and slaves leave less obvious traces than bronze, gold and amber.

In contrast, the images engraved on rocks by Bronze Age people are still there several millennia later.

***

Johan Ling wrote his doctoral thesis on Bronze Age rock carvings. Such engravings have been found at many places worldwide. But the largest concentration, including some of the largest and most striking images, is to be found in the municipality of Tanum, in Bohuslän. This is the site of

Vitlycke Museum, and the whole area around it has been declared a UNESCO World Heritage Site.

What was new in Ling's thesis was that he systematically compared rock art locations with geologists' estimates of the position of the coastline in earlier times. This reveals a pattern that is far from obvious to a person walking around among fields, grazing land and copses in the farming landscape of Bohuslän.

Today's Vitlycke lies several kilometres from the coast, a long way from seaside resorts such as Hamburgsund, Grebbestad and Fjällbacka. But the sea level during the Bronze Age was 15 metres (50 feet) higher than it is today. At the time when the Bronze Age people were engraving their rock art, the rocks were right on the shoreline. Vitlycke, for example, lay some way up a large sea inlet. This immediately makes it clearer why boats are such common motifs in these pictures. There are thousands of examples.

In the past, many archaeologists focused more on the farming landscape. They took it for granted that the people who engraved the images were primarily farmers. This led them to regard the boats as essentially symbolic, an element of Bronze Age religion. But we now know they were also an important part of everyday life.

Real boats presumably looked much like the ones depicted in the rock art. Although these are often highly stylised, their shape is clearly reminiscent of the Hjortspring boat. Prow and stern alike are elongated into points. The prows are sometimes adorned with an animal head. In many cases there are a series of short lines sticking up out of the boat, representing oarsmen. The large petroglyph at Vitlycke, just a few hundred metres from the museum, includes an image in which a number of rowers, all clearly distinguishable, are raising their oars in greeting. Lur players and acrobats performing backflips can be seen on other boats. There is absolutely no doubt about the existence of bronze lurs, which are a type of long blowing horn; 60 or so have been found, particularly in Denmark and southern Sweden, many having been placed in lakes and bogs. In most cases they were used in pairs. Some can still be played, giving us an inkling of how Bronze Age music may have sounded.

The most celebrated image on the Vitlycke rock represents a man and woman kissing. Their sexual organs are also joined. A common interpretation is that the couple in the picture are deities engaged in 'the holy wedding'. Such ritual weddings and fertility rites are known from much of Asia and Europe, from India to Iceland. In Greek mythology, it is Zeus and Hera who are thus united. In the *Poetic Edda*, the servant Skírnir describes how the god of fertility, Freyr, woos the giant (*jötunn*) Gerðr, who eventually yields. The *Edda* text is written like a play, according to modern experts in literature. One can picture the culmination of the final act in the trysting grove, with Freyr and Gerðr in the main roles.

Even more often than boats, the rock engravings feature what are known as cup marks – small, round dimples in the rock. Archaeologists believe such marks are particularly often associated with burial places. They are often linked with death, burial and rebirth. Cup marks can thus also be viewed as a kind of fertility symbol. And there is some actual evidence for that interpretation in Swedish history, according to an anecdote from Tisselskog, Dalsland, from the years of poor harvests in the 1860s. The farmers went to church and implored God to help them – in vain. But when the harvests continued to fail, they arranged for a young man and woman to have intercourse, in public, on an ancient rock pocked with cup marks. The young man's semen was collected and mixed with seed in one of the cups.

A more peculiar type of fertility rite is intercourse between people and animals. This, too, can be seen in the Vitlycke rock carving, which features a scene where a man is having sex with a horse. Here, too, there are clear parallels with ancient myths that recur throughout the Indo-European language area. Sex between man and beast is a variant of the holy wedding; the mare is actually an avatar of the fertility goddess.

Vedic scriptures from ancient India – contemporaneous with the Swedish Bronze Age – tell of a ritual called the

Ashvamedha. This involved making white stallions compete with each other, after which the victor was throttled. Strangling the horse was believed to be more honourable than cutting it with sharp implements. The dead horse was placed on a bed under a capacious cover. While the corpse was still warm, the queen crept under the cover and feigned intercourse by placing the stallion's penis between her thighs.

As late as the twelfth century, the monk Gerald of Wales describes a pagan rite that he observed in Ireland. The local clan was to choose a chieftain. The man selected as chieftain was obliged – according to Gerald – to have intercourse with a white mare first, in front of the whole clan. The mare was then killed, cut into pieces and cooked in a cauldron. The new chieftain was obliged to bathe in the cauldron and eat some of the meat. The rest of the meat was shared out among the onlookers.

Incidentally, it is interesting that white horses appear to have held such significance for Bronze Age people. Leif Andersson, a geneticist from Uppsala working on domestic animals, has identified a gene borne by horses whose coat greys over the years, eventually turning completely white. He has clear evidence that people began to breed horses early to accentuate that particular trait.

There is some uncertainty about when domesticated horses first arrived in western and northern Europe. It is unclear whether there were any horses at all in Sweden and Denmark during the Battleaxe period. However, horses were definitely present in the Bronze Age, and they were very important, not least in mythology. As mentioned earlier, the oldest known vestiges of horse-drawn wagons with two wheels were found at the River Tobol in western Siberia. They are approximately 4,000 years old. But it was not long before very similar vehicles reappeared in the Bohuslän rock engravings.

The Vitlycke rock engraving includes a man driving such a two-wheeled chariot, clearly drawn by a horse. The man holds the horse's reins in one hand and a hammer in the other. Before him, a flash of lightning can be seen. The most likely interpretation is that the man is a thunder god – the Bronze Age counterpart of the god later known to the Vikings as Thor. The thunder god's attribute was an axe or a hammer, with which he would strike to produce thunder and lightning.

The Celts called the god of thunder Taranis, while the Romans called him Jupiter and the Greeks Zeus. But everything suggests that the god of thunder and weather is far older, and that he was already known to the early Indo-Europeans. According to a reconstruction by linguists, the Indo-Europeans' sky god was called something like 'Diyéus' – quite similar to the Russian word for 'day', день (pronounced 'dyen'), and the French for 'God', Dieu. In Lithuanian, the sky god is called 'Dievas'. The early form of this Indo-European god seems to have been a father figure responsible for the sky and the day.

There was also a divine sun. Often, the sun was depicted as a wheel. The ancient Vedic scriptures of India contain many such references, which also occur in Greek plays and the *Poetic Edda* from Iceland. The Swedish rock engravings include many images showing how the sun moves during the day and at night, often borne by a ship.

Another common motif shows the sun riding in a horse-drawn vehicle. The most spectacular example is the Trundholm sun chariot, now displayed at the Danish National Museum in Copenhagen. It was found in a bog and probably dates back to the early Bronze Age. A horse cast in bronze stands on a little chariot. It is pulling a huge sun in the shape of two bronze discs, one side of which is gilded. One interpretation is that the gilded side represents day, while the disc made only of darker bronze symbolises night.

There are also images showing horses drawing the sun forth during the day, when it is light, while it is carried through the underworld by boat at night.

My thoughts turn to the Sun Salutation, which I learned on a yoga course a few years ago and still perform several times a week. It is a good way to stretch the muscles in your back and legs and strengthen your shoulders and arms. When

I was on the course at a yoga institute in Stockholm, I found the religious aspects of the teaching quite irksome. Before we started our yoga practice, we were supposed to recite verses from ancient Vedic scriptures, and the instructors made a big thing out of the fact that we were learning to perform movements going back several thousands of years. Personally, I was only interested in the physical exercise, especially balance and suppleness, aspects that yoga emphasises more than modern Western exercise programmes. Now, reading the literature on early Indo-European myths, I realise that the Sun Salutation may well have the same origins as the Trundholm chariot and the suns in the rock carvings.

The acrobats of the Bronze Age must have put a great deal of effort into improving their balance, strength and suppleness. Tiny bronze statuettes of acrobats performing backflips, like those in the rock carvings, have been found in Denmark. To judge by their appearance, there were both male and female acrobats. The female ones wore a special kind of short skirt made of woollen cords and clinking bits of bronze.

These acrobats probably played a set role in rites and provided entertainment at great feasts held by Bronze Age rulers to consolidate their power. Just as important were the poets – or the bards, to use the Celtic word. Today, poetry is often viewed as something elevated above the rest of society. But in the Bronze Age, poets were an essential part of the apparatus of power. They played a kind of PR role. It was the court poet's job to sing the praises of his lord and master as eloquently and convincingly as possible.

Two important objects used at ritual feasts out of doors were camping chairs and drinking vessels. Pictures of folding chairs have been found from Egypt to Scandinavia, and physical remains have been found in Danish bogs. Drinking vessels were sometimes made of bronze, rarely of glass, but more often of finely polished and decorated pottery. It seems clear that they contained alcohol.

And there is a fair bit of chemical evidence of what Bronze Age rulers used to raise a toast when sitting on their folding chairs, listening to poets and lur players and watching acrobats perform. The US anthropologist Patrick McGovern, the world's leading expert on the history of alcoholic beverages, has analysed shards of drinking vessels from Denmark and Gotland. He has come up with a recipe for what he calls 'Nordic grog'.

His research covered grave finds from four sites, the oldest from the early Bronze Age, the most recent (on Gotland) from the beginning of the Iron Age. All the beverages contained honey, produced by bees from flowers including lime tree blossom, dropwort, white clover and heather. Some drinks also contained vestiges of barley or wheat, as well as traces of cranberries, lingonberries, bog myrtle, yarrow, juniper berries, and spruce and birch resin. In two cases – the older of which is 3,100 years old – traces of wine made from grapes can also be detected.

McGovern's conclusion is that pure mead brewed using only honey was a status symbol drunk only by the highest-ranking people. However, Bronze Age people often mixed that exclusive honey with barley or wheat. In other words, they brewed a mixture of mead and beer, to which they added herbs and berries. 'Nordic grog' was also spiced up with wine on occasion. This wine was probably imported from the south along trade routes used for amber and other wares. The wine mixed into these beverages is yet another piece of evidence of the extensive networks that already existed in Europe by the Bronze Age.

For me, there is no doubt that boats and metal were two of the main driving forces behind the growth of Bronze Age networks within Europe. But the book by US archaeologist David Anthony is called *The Horse, the Wheel, and Language*. He chooses to emphasise the role played by domesticated horses in dispersing Indo-European languages and the Bronze Age lifestyle throughout Europe. Apart from trading in metal, this lifestyle included a new mythology, a heavily stratified class-based society and woollen clothing. Large flocks of sheep were part and parcel of the pastoralist culture, which spread from the steppes in the east. People began to wear clothing made of wool, rather than just of hides and plant fibres. The custom of showing group allegiance through particular colours and checked patterns may even have arisen at this early stage. This thought occurs to me on observing the marketing of tartans by Scottish-themed boutiques and firms like Burberry today.

David Anthony may well be right that horses played a decisive role in historical development. Admittedly, there are only a few finds of horses that show signs of having been domesticated from the Corded Ware period in central and northern Europe. There are rather more from the Bell Beaker period. But a little later, once the Bronze Age was well under way, horses became much more common. No doubt it was horses that took the Indo-European culture eastward and southward, for example to India.

Burials of people belonging to haplogroup R1a have been found in Kazakhstan and Krasnoyarsk, Siberia. These are between 3,800 and 2,300 years old and appear to represent the equestrian culture of the steppes on its way to the east.

Multiple factors were involved when the new networks and the new class system took over in Europe and parts of Asia: wheels, wagons, boats, bronze, woollen cloth, Indo-European languages, horses and amber. And we should not forget that they included trading in slaves.

CHAPTER THIRTY-TWO

Iron and the Plague

The arrival of iron was a serious blow to the networks of the Bronze Age. Conditions changed. Control over rare copper and tin mines no longer conferred the major advantage that it once had.

The raw materials for producing iron were available in many places. The nearest lake often yielded bog ore. Iron has a higher melting point than bronze. But people were quick to learn how to use bellows to pump in air during the smelting process. After that, bronze no longer had the same importance. Although iron was perhaps less attractive than golden-yellow bronze, it could be forged more easily and was more useful for tools and weapons alike.

In the Mediterranean region, the Iron Age culture emerged during the Roman era. Roman rule was to expand, taking over much of Europe and becoming one of the most powerful and influential cultures the world has ever seen. Of course, the Roman Empire and iron also affected the Nordic peoples. However, in many ways life there continued much as before, at least in southern Scandinavia.

The fall of the Western Roman Empire about 1,600 years ago ushered in that part of the Iron Age we know as the time of the great migrations or *Völkerwanderung*. The groups concerned migrated in different directions, and traces of their migrations can often be detected in the DNA of modern populations. This is particularly marked in certain regions, notably Great Britain and eastern Europe. Germanic tribes migrated from Denmark and north-west Germany to England. They came to be so dominant that the Anglo-Saxon dialects they spoke displaced longer-established Celtic languages, forming the basis for English. Groups speaking Slavic languages dispersed in all directions from their heartland between the Dnieper and the Dniester. The Sami languages probably came from the east to the areas where they are now spoken, and this was probably linked with the use of iron.

However, as I have tried to explain in this book, migrations of ethnic groups are not a phenomenon confined to that specific period. Such migrations have occurred ever since we anatomically modern people first arrived in Europe, over 40,000 years ago.

According to the latest DNA research, the population of today's Europe bears the stamp of three great waves of migration, above all: hunters who came here during the Ice Age, farmers who came from the Middle East bringing the earliest agriculture, and a third wave from the steppes in the east, who brought Indo-European languages with them.

***

During the Iron Age, the climate became somewhat cooler than it had been during the Bronze Age. A number of historical sources speak of the 'Fimbulwinter', which seems to have arrived around AD 536. This was a period without summers, during which the harvests failed for several consecutive years. Texts such as Snorri Sturluson's *Edda* from Iceland and the Finnish *Kalevala* describe how the sun could barely be glimpsed behind the clouds. And there is scientific evidence of several unusually cold years around AD 536 – in fact, it was the most severe period of cold for several thousand years. Tree rings show a virtual hiatus in growth for several seasons around that time. Greenland ice cores reveal traces of particularly high sulphuric acid levels, pointing to a volcanic eruption.

As luck would have it, I chanced to meet Michael Baillie, the retired professor of geology who shares a small study at Queen's University, Belfast, with James Mallory, the expert in the origins of the Indo-European languages. Baillie told me it was he who first noticed that annual growth rings provided evidence of a period of extreme cold around AD 536. He published articles on this back in the 1980s and still has a strong interest in the subject.

The dominant theory is that the Fimbulwinter was caused by a huge volcanic eruption that may have taken place in today's El Salvador. Baillie believes that several discrete events took place more or less simultaneously, at least one of them probably being the impact of a comet. The details are somewhat vague, but scientists are beginning to fill out the ancient tales with facts, revealing what actually happened to the summers after AD 536. What is clear is that they were far colder than usual for a number of years.

More data is also emerging about a great epidemic that raged immediately after the Fimbulwinter and had a radical impact on our history – an early outbreak of the plague.

The plague used to be regarded more as a mediaeval scourge that struck hardest in the fourteenth century. The plague epidemics that ravaged Europe in the Middle Ages are described in detail in a number of historical sources. The first wave is usually known as the Great Mortality, the Great Plague or the Black Death. It reached the Crimea in 1346 and spread through Europe in the years that followed, killing about one person in every two. The disease ran its course rapidly; most of those infected were dead within a matter of days. They suffered from headaches and high fever, and their lymph nodes swelled up into great boils or buboes, hence the name 'bubonic plague'. In many cases this was accompanied by subcutaneous haemorrhages that turned the skin dark blue. Sometimes the bacteria affected the lungs, causing pneumonic plague. This variant has a mortality rate of nearly 100% and is highly infectious.

The epidemic continued to break out in waves over several centuries, during which it gradually became less deadly. Over time, people's ability to withstand the plague grew. They learned how to deal with infection by such means as quarantine. This was also a biological process; over many generations, those whose immune system was least well adapted to combating plague microbes died, while those whose immune system happened to be specifically resistant to them survived and had children. This adaptation has left lasting traces in Europeans' genetic material.

In the past, researchers questioned whether the Black Death really hit Sweden as hard as the plague-ridden countries of southern Europe. But historians have now been able to confirm that it did. Janken Myrdal, for instance, an expert in agrarian history, has compiled an extensive range of evidence including letters, invoices and wills, abandoned farms and newly built houses. One particularly striking piece of evidence is a report from an area of Skåne where virtually all ecclesiastical art ceased for nearly a century. That is testament to how the Black Death scourged society at many different levels.

Some researchers have questioned whether the Black Death of the fourteenth century was really caused by the plague bacterium, *Yersinia pestis*, which was described in the nineteenth century and continues to cause outbreaks in Asia and Africa. They have suggested all kinds of alternatives, including various viral diseases. However, in recent years DNA researchers have been able to show clearly that the plague microbe was the cause of the Black Death. Two research teams have examined victims found in mass graves dating from the years around 1350. They have exhumed remains in churchyards in Italy, France, Germany, the Netherlands and England. Both competing research teams have succeeded in isolating traces of the plague bacterium. Thanks to DNA analysis, they have also been able to reconstruct a genealogical tree with different strains of the plague microbe, enabling them to confirm that the disease originated in Asia, as reported in a number of historical sources.

No one can now deny that the plague bacterium really was a terrible scourge in mediaeval Europe. And it looks as if similar outbreaks of disease struck Europe far earlier, way back in the Iron Age.

Around AD 540 – just a few years after the long Fimbulwinter – the historian Procopius described a major outbreak of the plague in Constantinople. This outbreak is usually referred to as the Justinian plague after Justinian I, the emperor of the eastern Roman Empire at that time. There are far fewer written sources from the sixth century than from the Middle Ages, and historians have been even more divided on the subject of this epidemic. There has been controversy over both the nature of the Justinian plague and whether it reached the lands north of the Alps. For a long time, it was conventional wisdom among Swedish researchers that the plague had not had any noticeable impact on Sweden during the Iron Age. But Per Lagerås, an archaeologist from Lund, has dated clearance cairns, the heaps of stones that farmers gathered together when breaking new ground in preparation for cultivation. He has also drilled down into the soil and carried out pollen analyses. This has enabled him to show that many farms in the uplands of Småland were abandoned at the time of the Fimbulwinter and the Justinian plague. His findings suggest a rapid decline in population.

And a few separate research teams have now been able to demonstrate that the Justinian plague really was *the* plague, caused by the plague bacterium *Yersinia pestis*, just as in the Middle Ages. However, the Justinian plague represented a separate branch on the bacterium's genealogical tree. DNA mapping shows that the disease jumped from animals to people on at least two separate occasions. One such transfer caused the Justinian plague in the sixth century, while the Black Death of the fourteenth century was caused by another transition from animals to humans.

One of the carriers was the black rat. This animal probably came to Europe from Asia at the time of the Romans and thrived in our granaries, which were often in lofts in family homes. The black rat's fur harboured fleas that passed the germs on from animals to people.

But black rats and their fleas are not the whole explanation. Other rodents can also act as hosts, and plague germs can also spread from one person to another. The strain known as pneumonic plague, in particular, is an airborne disease, which can easily be spread without any animal involvement.

The churchyard where researchers can now identify victims of the sixth-century Justinian plague is in Germany. This means no one can now claim that that outbreak was

confined to the lands south of the Alps. Just like the Black
Death, the Justinian plague had a major impact on Europe's
development. It probably killed a large proportion of the
population, especially as people's immune systems were
completely unprepared and the societies in which they lived
were already vulnerable, owing to the failed harvests of the
Fimbulwinter.

Clearly, the population gradually recovered from the Black
Death of the fourteenth century. Per Lagerås's view is that
these outbreaks of disease, seen with the benefit of hindsight,
were not an unmitigated disaster in every respect. He
describes them as dealing 'the coup de grâce to feudalism', the
social system under which the poor worked like slaves, under
slave-like conditions, for richer farmers and landowners.
According to Lagerås, outbreaks of plague were followed by
labour shortages, obliging landowners to pay better wages
and provide improved working conditions.

Just a few decades after the Fimbulwinter that began in
536 and the Justinian plague of 540 the population began to
recover. A few centuries later came the last stage of the Iron
Age in Scandinavia, which we know as the Viking Age. In
many respects, the mores of this time were reminiscent of the
Indo-European lifestyle that had already arrived in Sweden
by Bronze Age times.

CHAPTER THIRTY-THREE

Am I a Viking?

In reading up on the Viking era, I am struck by a palpable contrast. Books by Scandinavian authors emphasise the attractive aspects: trade, wood carving, iron-working, the art of boat-building, poetry, feasts, mead, costume, jewellery and, in more recent times, women of high status who wielded a good deal of power. Books by British authors have a different angle, tending to focus more on terror, raids, blackmail, abduction and slave trading.

This gives me pause for thought. During my years as a science journalist, most of my reporting has focused on research in the natural sciences. Results in these disciplines rarely differ according to scientists' country of origin. I am reminded that history and archaeology have developed in close connection with ideological and national interests. The link between nationalism and historical research was much stronger in the nineteenth century and the first half of the twentieth. In some parts of the world, archaeologists and historians still face political pressure to purvey conclusions favourable to the interests of the powerful. But even researchers who are fortunate enough to be able to work freely and independently of political authorities assess information in different ways, depending on their own point of view.

All the books I read on the Viking era and the exhibitions I visit are of a very high standard. Authors and curators alike take a scholarly approach and reference their sources. It's just that they have a slightly different angle and focus. Such differences, though small, are enough to convey radically different views of the Viking Age, one being the aggressors' view of events, the other the victims'.

What I want to know is this – which side were my own ancestors on?

***

The dual perspective begins with the word 'Viking' itself, which can mean several different things. Initially, it referred to a phenomenon, not a person. The Old Norse expression '*fara í víking*' ('to go on a Viking voyage') meant to engage in piracy – to take ship, sail to a place some distance away, together with a number of companions, and grab whatever you could, be it valuables or the human spoils of war.

The word is also used, including by professional historians, in the expression 'the Viking Age', which refers to the last part of the Iron Age. A common definition is that the Viking Age began with the raid on the monastery of Lindisfarne (Holy Island) in AD 793, and ended by about 1100, by which time Christianity had become the dominant religion in Scandinavia.

To confuse matters further, the word 'Viking' is often used in a very general sense to describe a person of Scandinavian origin living at the time of the Vikings. The stereotype is a red-haired, bearded strongman quaffing mead from a large cow horn and sporting a horned helmet. This cliché is common in popular culture; it often turns up in advertising and among supporters at international matches.

Professional historians have a somewhat different view of the Vikings. The notion that they had horned helmets is a complete misconception. Most of the population of Scandinavia at the time of the Vikings lived on farming, hunting and fishing, not from piracy.

The attack on the Lindisfarne monastery has come to mark the beginning of the Viking Age, mainly because it is described in such detail in writing. The ecclesiastic Alcuin described the raid in a poem and several letters. He tells of how 'the pagans' desecrated God's sanctuary, how blood ran around the altar and how the bodies of saints were 'trampled [...] like dung in the street'. In all probability, there had been violent raids from Scandinavian shores in earlier times, but they were not documented in any written sources preserved for posterity.

There have been trade and exchanges between the British Isles, the Atlantic coast of Europe and the Scandinavian lands at least since the Bronze Age. Such encounters could be violent. The Danish Hjortspring boat from about 350 BC was quite clearly built for war, given its symmetric design and the various weapons and shields found nearby. There have been sophisticated rowing vessels ever since the Bronze Age. It was nothing new for dozens of athletic men to be able to row long distances, but the Vikings were also able to sail. They learned the art late by comparison with Mediterranean seafarers. However, once the Scandinavians started to combine their well-crafted clinker-built vessels with their oarsmanship and woollen or linen sails, they had the upper hand for a while. Their tactic was to strike from the water, and they were smart enough to choose victims who were unable to put up any armed resistance.

Moreover, the Vikings were not inhibited by a taboo that placed restrictions on other European warlords. Not being Christians – at least not to begin with – they had no qualms about raiding monasteries and churches.

Swedish-born Anders Winroth of Yale University claims in a new book that the Vikings were not really any more brutal than other contemporary European warlords, such as Charlemagne in France. The difference was that Vikings often raided monasteries and churches, inhabited by priests and monks who had mastered the art of writing. This means their depredations are unusually well documented, at least as regards western Europe. Consequently, we know a good deal about the brutality of bands of Scandinavian pirates who attacked monasteries, churches, farms and towns in the British Isles, along the Atlantic coast of Europe and on the major rivers of France.

Conversely, historians are less well informed about the voyages to the east, to regions where literate people were more of a rarity. The best-documented encounters are those with Arab writers. Findings from historical and archaeological sources show that there was extensive trade in the east as well. The Scandinavians traded mainly in furs and slaves. They purchased – and plundered – all sorts of luxury goods, including silver and silk. Their networks extended as far as central Asia.

Early on in the Viking era, Scandinavians also began to stay on and settle down in new places. Sometimes local rulers encouraged them, a settlement of armed Scandinavians being viewed as a good defence against raids by other Vikings. Scandinavians colonised Iceland and, for a while, even Greenland, where walruses with valuable ivory tusks were plentiful. From Greenland, they undertook expeditions to Canada to fetch timber. It is very likely that they also took women.

Voyages to regions that the Icelanders called Helluland, Markland and Vinland are described in mediaeval sagas. Archaeologists have confirmed that these places may have been in North America. The remains of a Viking-style settlement have been found at L'Anse aux Meadows on Newfoundland. Geneticists, too, are now helping to discover what happened when Scandinavians visited America – several hundred years before Christopher Columbus strode ashore.

A number of today's Icelanders bear in their DNA the evidence of a voyage to America. These people have mitochondria belonging to haplogroup C1 and are descended from the same woman. Calculations show that she came to Iceland around AD 1000, very probably from America. Haplogroup C1 occurs almost exclusively among the indigenous population of America. The woman's living descendants in Iceland belong to a subgroup that has not been found in modern-day America. Either the researchers have missed it in their analyses, or the whole group has died out everywhere except Iceland. Admittedly, one might come up with other, more tortuous interpretations, such as that haplogroup C exists in Siberia – and could thus have occurred in individual Scandinavian settlers. But the most reasonable explanation is that the woman sailed to Iceland from Canada, via Greenland.

Most male settlers came to Iceland from Scandinavia – mainly Norway – in the early ninth century. This is clear from written sources, linguistic research and modern Icelandic men's Y-chromosome DNA. About 80 per cent of modern Icelandic men's Y chromosomes indicate Scandinavian origins. But it seems likely that the remaining 20 per cent come from Ireland. Icelandic researchers see a link between this part of genetic make-up and mediaeval sagas that tell how Icelanders took Gaelic slaves from the lands that are now Scotland and Ireland.

Many such slaves were women, according to the written sources. This is supported by genetic evidence. The Gaelic input in modern Icelanders' maternal lineage is even greater than that in their paternal lineage. More than half of today's Icelanders are descended from women with roots in Scotland and Ireland. That is clear from their mitochondrial DNA.

Of course, Viking Age Scandinavian colonies across Europe also included some women who themselves came from Scandinavia. Some feature in the mediaeval sagas, and some of them were both rich and powerful. New DNA studies reveal a pattern. The nearer to Scandinavia the colonies were, the higher the proportion of Scandinavian women. In the case of the Shetland and Orkney Islands, entire families seem to have sailed from Norway, with their domestic animals, in order to settle and start a life as farmers. We can tell this because both Y-chromosomal and mitochondrial DNA show a large proportion of Norse ancestry.

However, the further away the Vikings moved from Scandinavia, the fewer women there were from their native land. The share of mitochondrial DNA of Scandinavian origin declines with increasing distance. Nearly half the people living on the Shetlands today have Scandinavian ancestry, and the proportion is roughly equal in the maternal and the paternal lines. Among the inhabitants of the Orkney Islands, about one third of forefathers and foremothers alike are from Scandinavia. On the Hebrides, however, Scandinavian forebears account for a smaller proportion of people's ancestry, and can be seen only in the paternal line. This means that most of the incoming Scandinavians were male Vikings.

The Vikings left a relatively small genetic footprint elsewhere in Britain, according to analyses of nuclear DNA. But there are some traces. The Y chromosomes of some modern-day British men show where the Vikings arrived and sowed their seed.

Some DNA research companies in Britain have been criticised for making too much out of people today being 'Vikings', just because they have some DNA from Scandinavia. Such 'Viking' claims are misleading, of course. The British men concerned are a mix of ancestors from different places, just like the rest of us. However, they are descended from Scandinavians in the paternal line – which may be interesting to know. When private individuals like these have their DNA tested, they help the rest of us to find out more about the voyages and the dealings of the Vikings.

One circumstance that complicates this research is the fact that there had already been a major wave of immigration to the British Isles from Denmark and Germany during the time of the Great Migrations, around AD 500. Some of these Anglo-Saxons came from roughly the same Germanic groups as some of the Vikings a few centuries later. This means very detailed DNA analyses are needed to enable researchers to distinguish the patterns of movement specific to the Viking Age. Earlier analyses, conducted by both academic researchers and firms providing services to family history researchers, were not sufficiently high resolution. But today the more sophisticated tests sold to private individuals have become sufficiently detailed.

As a result, amateur genealogy researchers are now constructing ever more detailed genealogical trees showing the travels of the Vikings. It can be clearly seen how men from Norway had sons in Scotland and northern England. The colonists in south-east England more often came from Denmark and southern Sweden – which is in line with what archaeological finds and documentary sources can tell us.

Private individuals researching their family history are also working on mapping the Vikings' travels in the east. One example is a special project involving men who believe they are descended from Rurik – the man who, according to written sources, was brought from Scandinavia in the ninth century to found the east Slavic state of Kievan Rus'. Research suggests that Rurik's Y chromosomes belonged to a particular subgroup of N3 that is particularly common in Finland.

Peter Sjölund, who uses DNA to research genealogy, has worked together with some of his Russian counterparts, identifying about 10 men in Russia and Ukraine who belong to a particular subgroup of I1a and have a paternal lineage that appears to originate with Scandinavians living in eleventh-century Kiev.

These findings support historical theories that Scandinavian Vikings – Rus or Varangians, as they were also known – played a key role in founding the first Russian realm.

Two haplogroups of Y-chromosome DNA are generally regarded as clear markers of Viking ancestry. One of them is I1a. The other is R1a – the very haplogroup I know my paternal grandfather Eric belonged to, thanks to Uncle Anders's DNA test. Might my forefathers have been Vikings who sailed eastward and westward, capturing slaves, trading in furs and bringing home treasures of silver? Might I, in fact, be a Viking of sorts?

***

My paternal grandfather, Eric Bojs, was a kindly, humorous man. Trained as a primary school teacher, he eventually went on to teach at a teacher training college in Kalmar. Alongside his work, he devoted himself to two absorbing interests in his free time.

One of them was radio. With a wind-up tape recorder on the back of his bike, he cycled around recording reports for the radio. For many years he single-handedly ran what was later to become Swedish Radio's Kalmar department. He interviewed Olof Palme as a newly appointed minister and the last soldier recruited under the Swedish 'allotment' system, and every year, on Christmas Eve, he reported on the weather in Kalmar for national radio. On a number of occasions he also took on the role of a reporter for Swedish Television. His items included an interview with the brother of a suspected pyromaniac on the neighbouring island of Öland.

But my grandfather's favourite pastime of all was drawing. During his military service he discovered his ability as a quick, deft draughtsman, and he was trained as a military artist. Eric illustrated the first two books by Vilhelm Moberg, the author of a classic series about Swedish emigrants to the United States. He later travelled around to community halls and other venues to entertain the public.

Eric also used his talent as a draughtsman to create maps, games and puzzles, and he pioneered visual pedagogy in Sweden. He also painted watercolours and, on occasion, large murals for public spaces. Most of his murals are to be found in Börjes, a low-cost department store in Tingsryd, a small town in the southern province of Småland. Eric lived there for a few years as a young, newly married primary teacher, and my father Göran was born there.

One summer's day I gather together my paternal uncle, a few cousins, my brother and one of his sons for a family reunion. We begin with coffee and a look at my grandfather's murals in the cafeteria in Börjes. The owner, one of the sons of the original Börjes, shows us around. Having Bojs as a surname is a particular advantage here in the Tingsryd area, as my grandfather was clearly a local celebrity and very popular. The style of the pictures on the walls of the department store can be described as rural romanticism. They represent scenes from rural life, from my grandfather's childhood and earlier centuries. Horses are the most frequent motif. Börjes started off by selling low-cost equipment for horses, and although it has now diversified, horses still have a central place.

During my childhood Tingsryd was best known for its brewery and its beer, and today one of the major employers is a boat factory. Beer, horses and boats – I can't help thinking that these form a particularly fitting combination for a family reunion taking us back to our roots. These, after all, were the three factors that drove agriculture and the Bronze Age, and which were instrumental in bringing our shared lineage to Småland.

Eric was born on a medium-sized farm in Väckelsång, about 10 kilometres (6¼ miles) north of Tingsryd. According to family lore, his grandfather – my great-great-grandfather – began working life as a poor railway worker. But he met a girl whose father was a lay judge, who helped the young couple to buy the farm.

In front of the church in Väckelsång stands a large model of a cow. It is an enlarged copy of a wooden toy that my grandfather made in his youth, during the short period when he and his brothers ran a toy factory together. According to family history, this cow came to be a symbol of the local dairy, which was eventually taken over by the dairy company Arla. So it is basically this cow that features on Arla's milk packaging today. To see the original, you need to go to Väckelsång.

From Väckelsång, we go on to Urshult, a few kilometres to the east. I have been helped by some family history researchers who were quick to discover that our paternal line actually came from that area – something we had no idea of.

Urshult has no obvious signs of Eric Bojs, such as statues of cows or mural paintings. What it does have, however, is a very active local history society that has been helpful in putting me in touch with a local family historian. This man, Klas Samuelsson, turns out to be related to my family in various ways, though not through the paternal line, which is what we are looking into on this occasion.

Klas Samuelsson leads us to a red farmhouse, Froaryd Södergård, which stands next to Lake Åsnen. A few cows belonging to the present owner are grazing along the shoreline. The grass is the intense green that comes only from many centuries of regular grazing. With the help of old court records, Klas Samuelsson has managed to find some information about our ancestor Måns Månsson, who was born here in the mid-seventeenth century. The family were originally tenant farmers who leased the farm from the Crown. In the eighteenth century Måns Månsson's descendants were able to buy the farmhouse and its land.

After some calculation, I work out that the youngest person in our group – my brother's son – represents the eleventh generation in the direct line after Måns Månsson.

In Måns Månsson's time the farmhouse was presumably grey rather than red, and the roof would have been covered in turf or thatched in reed, rather than tiled. In other respects it probably looked much the same. Apple trees in the meadows sloping down towards the lake would have been covered in blossom in spring. Today, the area around the south side of Lake Åsnen remains famous for its apples, grown in meadowlands in the traditional way. Some growers actually receive EU grants for maintaining this old-fashioned type of cultivation, which is not particularly productive in terms of the apples' economic value. But the meadows enable biodiversity to thrive – with a wealth of flowers, including many orchids – and they are viewed as very valuable in terms of cultural history.

Unfortunately, things went downhill for our ancestors a few generations after Måns Månsson. His son's son, Per Johansson, moved to a farm a few kilometres to the north, whose location is not at all as attractive or close to the water. The farmyard is messy, and a few hostile dogs bark at us. We have been warned that the current tenants have a criminal background.

Per Johansson's son, Johan Persson, also seems to have let things slide when he lived here in the late eighteenth century. Hard liquor may have had a hand in that. According to the notes taken by the pastor after his *husförhör*, a yearly visit to the farm to check on the inhabitants' literacy and knowledge of the Bible and the Lutheran catechism, Johan Persson had 'completely forgotten his Christianity'. The archives record a spell in the fortress at Karlshamn. Johan and his family were forced to leave the farm, after which they moved around from farm to farm within the area. Johan's son, Nils, ended his days at the age of 59 in the poorhouse at Tävelsås. But it was that same Nils who fathered Peter Nilsson – the railway worker whose father-in-law helped him to buy the farm in Väckelsång where my grandfather Eric Bojs was born.

My relatives and I had no idea there had been such ups and downs – deep downs – in our forebears' history. We are a little overwhelmed when we take a seat in the yard outside Urshult's open-air museum to digest what we have just heard. There have been turbulent swings in the family's fortunes, with social mobility in both directions. But the location has remained more or less the same. Apart from a few minor moves, the family clearly remained within a radius of a few dozen kilometres – at any rate between the birth of Måns Månsson in Urshult in the 1640s and that of my father, Göran, in Tingsryd in 1931.

And when I consult Peter Sjölund, the genealogy researcher who drew up the extensive genealogical tree showing the descendants of 'Ragnar', it turns out that our ancestors scarcely moved in the course of 4,000 years. They stayed at home when they had sons, in any event.

The genealogical tree of men belonging to haplogroup R1a-Z284 has several branches, three of which are particularly prominent. One of these seems to have diverged early on from the west coast of Sweden towards Norway. This branch includes many of the men who became Vikings along the Atlantic coastline of Europe. The second branch shifted to the Mälaren Valley. But the third of these main branches went towards southern Sweden. Many of its offshoots remained on the southern side of the high plain of Småland for thousands of years.

Even the expert DNA genealogist Peter Sjölund thinks that sounds odd, bearing in mind how people have travelled around in the course of history. But I get to thinking about a family reunion in Väckelsång's community hall that I attended when my paternal grandfather and grandmother were still alive. All of my grandfather's siblings were there. And it struck me that my grandfather was the only one in a big family to have left the region around Lake Åsnen. He was the most adventurous of them, moving away as far as Kalmar. A generation later, his children moved outside the confines of Småland to study in the university cities of Lund, Stockholm and Uppsala. And later on they ended up in Lund, Gothenburg and Kristianstad.

The vast majority of my forebears and relatives stayed in the area where they were born. That, after all, is a fairly ordinary pattern. However, things look different on my paternal grandmother's side.

***

I explained earlier how the first farmers in Europe migrated from Syria, via Greece and the Balkans, to central and northern Europe. And I showed how the distribution of people with the same mutations in their mitochondrial DNA as my paternal grandmother reflects those migrations almost perfectly.

When the DNA of some Stone Age farmers from Västergötland was published in *Science*, putting them in the global limelight, I experienced a small personal thrill. These farmers were exhumed only a few dozen kilometres away from the place near Falköping where my paternal grandmother's foremothers lived in the eighteenth century. It felt almost as if the researchers had exhumed members of my family and carried out DNA tests on them.

In my memories of my childhood and youth, there is a very special aura surrounding my paternal grandmother, Hilda. In appearance, at least, she might well have been one of Europe's earliest farmers. She had dark hair and brown eyes, and with her fine features, she was strikingly beautiful. She also had a fine singing voice; she would recount with pride how she used to sing solos in church as a child. Her father was a cantor at the church in Öjaby near Växjö. He was strict, with numerous children and a low income. One day a wealthy couple from Stockholm came by to ask whether they could take the pretty little girl home with them for a consideration. But the cantor was furious; poor though he might be, he had absolutely no intention of selling his children. I never really established whether my grandmother was happy to be able to stay with her family and her strict cantor of a father, or whether she fantasised about the turn life might have taken in a well-to-do Stockholm family.

The children of Öjaby were only able to attend school for half days, as the schoolteacher also had to cycle over to the neighbouring parish of Härlöv to teach there. But little Hilda was allowed to ride on the back of the teacher's bike and go to school in Härlöv too. After an excruciating year as a downtrodden nursemaid in an officer's family in Stockholm, she managed to train as an elementary teacher at the teacher training college in Växjö, where she met my grandfather. As the mother of two children she proposed to my grandfather that she provide for the whole family through her work as a teacher. That would have enabled him to throw himself fully into training as an artist. He had taken a preparatory course, and several of the participants had subsequently become well-known artists. But my grandfather turned down my grandmother's offer to take over as the breadwinner. He never became a famous oil painter, but continued his long career as a teacher and amateur artist. Grandmother Hilda stopped working a few years later and remained a housewife for the rest of her life.

When I was six years old and about to start school, she travelled over from Kalmar to Gothenburg for my first day. My little brother was a baby at the time, and my mother needed help. I also recall some magical Christmases in Kalmar and a number of Easter weekends. The blue anemones made a particularly big impression on me, and so did the sago pudding in my grandmother's kitchen. There were no blue anemones and no sago pudding at home in Gothenburg. When I was 12 or so she taught me how to darn socks. We repaired the runs in my nylons together. Thrift was deeply imprinted in her. When you peeled potatoes, the peel should be as thin as you could possibly manage, she taught me.

After the age of 12, it was many years until I saw my paternal grandparents again. My grandmother never forgot to send Christmas and birthday presents – mostly clothes, sometimes simple jewellery. Her choice of gifts always showed perfect taste and great thoughtfulness.

As soon as I was on my way to adulthood, I got in touch with my paternal grandparents again of my own accord, after which I visited them in Kalmar several times. My grandmother continued to live a frugal, thrifty life well into her eighties. She would go on long walks, pick mushrooms and cycle over the new bridge all the way to Öland to look at the blue anemones there.

Now I do not imagine for a moment that the genetic material which determined my grandmother's appearance remained unadulterated all the way from Syria to Småland. It is quite clear to me that she was a mix of Stone Age farmers and hunters just like all of us, and that a great deal has happened since the first farmers left Syria nearly 10,000 years ago. One need only go back two generations to see that my grandmother was descended from four different people. A few generations further back, and her forebears can be reckoned in the thousands. What traits are passed on is the result of chance.

But all this has to do with feelings as well. For me, there is a particular symbolism in the fact that mothers and daughters followed each other in a long, unbroken line for thousands of years and that it is now possible, thanks to DNA technology and mitochondria, to see what paths they took.

There are about 400 generations between me and the women farmers in the ancient Syrian graves – those who also belonged to haplogroup H, whom, for simplicity's sake, we can call 'Helena's clan'. Eight generations separate me from the oldest documented female ancestor on my paternal grandmother's side. She was called Katarina Eriksdotter and was married to a shoemaker called Petter Andersson. They and their children lived in the little village of Storskogen in the parish of Dala, about 10 kilometres (6¼ miles) north of Falköping.

I take advantage of my trip to Falköping – to interview the archaeologist who excavated the Gökhem farmers – to fit in a visit to Dala. The house in Storskogen where Katarina Eriksdotter once lived no longer exists. The old village of Storskogen is now a dense fir plantation. However, the rolling green landscape that surrounds it looks as it must have done in the eighteenth century, with plenty of old oaks and other deciduous trees, and cows grazing in the fields.

Unlike the old village, the manor house to which the shoemaker's home belonged has survived. It still stands right next to the church, with which it is linked by a private passageway. 'A good illustration of the proximity of temporal power and Church,' observes my local informant, a family history researcher and a former member of the local council for the liberal agrarian Centre Party.

The owner of the Stora Dala estate in Katarina Eriksdotter's day was one Peter Tham, whose family had become wealthy thanks to the East India Company. The estate is supposed to have been one of the largest and richest in the whole of Västergötland, which is saying something, as this is a particularly fertile farming area.

The shoemaker's small homestead in Storskogen also seems to have been relatively prosperous. According to the list of personal effects drawn up on Katarina's death, she owned two gowns of grogram, one black, the other blue; a brown satin jacket; a cloth jacket lined with leather; striped stays; a mohair cap; a silken cap; and several other garments. The farm included four red cows, a brown mare, and several sheep and pigs. The house contained looking glasses, a wall clock, brandy pans, liquor glasses and silver goblets; moreover, Katarina and Petter owned assets amounting to around 100 Swedish *riksdaler*.

According to the earliest traces of Katarina Eriksdotter in the church records, she moved to the parish of Dala to work as a maid on Häggestorp farm in 1766. She was apparently 26 years old at the time. The shoemaker Petter Andersson moved to Häggestorp in the same year, and the two married the year after. The marriage book states that Katarina was born in Odensåker, over 40 kilometres (25 miles) north of Dala. For me, 40 kilometres, or 25 miles, sounds like a long way for a young woman to move in the eighteenth century. I imagine that the Tham family's large farm provided employment opportunities that were lacking at home in Odensåker.

According to Dala's church records, Katarina was born around 1739, possibly in 1740 or 1741. Her father was apparently called Erik Jonsson. But no such birth seems to have been recorded in Odensåker. All traces of Katarina end there.

The person who helped me to research Katarina's background is a fellow student from my journalism course, Håkan Skogsjö. These days he is a writer, publisher and local historian in the Finland–Swedish archipelago of Åland. He has been researching family history since his teens and is one of the foremost researchers in the field throughout the Swedish-speaking area. If Håkan Skogsjö cannot identify Katarina's childhood home, that means it must be extremely hard to find her, and maybe just impossible. I also write to the local family history association in Mariestad, the municipality Odensåker belongs to, but there is no information to be had there either.

My only clue for the time before 1766 is DNA – and that proves baffling as well. Not a single person with absolutely identical mitochondrial DNA is registered in the databases where family historians record their results.

If you go back thousands of years, on the other hand, my paternal grandmother and I have huge numbers of relatives. Parts of their mitochondria have the same DNA set. Many people have put on record that they belong to the particular branch of the genealogical tree beginning with 'Helena' (a woman from the group of early farmers) known as haplogroup H1g1. They can trace their historical origins back to Greece, Albania, Croatia, Serbia, Hungary, Germany and Belgium, and, on the other side of the English Channel, to England and Scotland.

Many other family history researchers find hundreds of people whose mitochondria have identical DNA sets. There is a perfect match, with no differences arising from mutations. But when I search for perfect matches with my paternal grandmother's DNA set, I cannot find a single one: not in Västergötland, not in Sweden, not in Europe – in fact, nowhere in the world. It looks as if my paternal grandmother's forebears differed from those of my paternal grandfather in *not* living and working within a small, confined area for thousands of years.

However, once I've toned down my requirements and started looking not for a perfect match, but for people with a DNA set that differs by a single mutation, some matches turn up. I find 20 test subjects with a partial match. Their earliest known foremothers are not from Sweden; nearly all are from Scotland.

If I lower my requirements still further to a difference of two mutations from a perfect match, I find more partial matches. But still none of them are from Västergötland or anywhere else in Sweden. Most of my relatives are clustered in Scotland and northern England.

Mutations occur entirely at random. On average, they become a permanent feature of a lineage once in a thousand years at most. So the DNA results suggest that my paternal grandmother and the Scottish matches had a common foremother who lived about a thousand years ago. She seems to have lived in Scotland. If this is so, my foremother appears to have been on her own in leaving Scotland and settling in Scandinavia.

I realise I am in speculative territory now, but I have a theory about how that could have come about. It's not an agreeable theory. In fact, it saddens me. But I suspect that my grandmother's foremother travelled from Scotland to southern Scandinavia on board a Viking ship. She may conceivably have come as the lawful spouse of a successful Viking. But she may also have been a slave woman – a victim of the human trafficking of the day. It is common knowledge that the Vikings engaged in slave trading. Along with furs, slaves were probably their main export product.

The Arab writer Ahmad Ibn Fadlan tells of a tenth-century burial among a group of traders on the Volga, whom he refers to as 'Rūs' or 'Rūsiyyah'. This group probably consisted at least partly of people from the eastern part of Sweden. In any event, Ibn Fadlan's text is frequently cited as one of the oldest written descriptions of Vikings. He describes their 'perfect physique', for instance; they are tall and stately like date palms, fair-haired and ruddy, and extensively tattooed. Their personal hygiene, however, leaves a great deal to be desired by Arab standards.

In the same text, Ibn Fadlan also describes how a slave girl is burned on a pyre together with her dead master. First the Rūsiyyah pour large quantities of alcohol down her throat. Several of the men have intercourse with her, one after the other, and just before the fire is lit she is subjected to gang rape. All the while, the onlooking men beat their shields hard, making a din to prevent other slave girls from hearing her screams. She is killed when an older woman, known as 'the Angel of Death', plunges a dagger into her breast, while two of the men pull a noose tight around her neck. When the girl is dead and everything is finished, they light the funeral pyre.

Should we be sceptical of Ibn Fadlan's account, in the name of source criticism? Some historians are, pointing out that it might have been in his interest to depict the Vikings as brutal and sexually degenerate. But there is also support for his account from various quarters. In Viking times it was common for the dead – particularly the wealthy and prominent – to be burned on ships. They were accompanied by rich grave gifts, including dogs, horses and, very probably, human sacrifices. In a number of cases, archaeologists have interpreted human remains on burnt ships as thralls (slaves) who had been killed. There are some particularly clear examples in graves on the Isle of Man, but Scandinavia, too, has a number of sites which suggest that thralls – together with dogs, horses and other animals – were put to death when their master died.

A Norwegian study of a number of double and triple burials where only one of the corpses in the grave still had its head was published recently. The other individuals appear to have been beheaded. Archaeologists already suspected that the beheaded individuals were thralls. Now, new DNA analyses confirm that they were of different genetic origin from the individuals who were interred with their heads. In addition, the individuals thought to be thralls had a different diet. Isotopes in bones and teeth indicate that the people who were buried intact had a more varied diet, including a good deal of meat from land animals. The thralls, on the other hand, ate mostly fish.

Dublin, now the capital of the Republic of Ireland, was originally built by Vikings, and was a major slave-trading market. In the Annals of Ulster, an unknown writer reported for the year AD 821 that 'Étar was plundered by the heathens, and they carried off a great number of women into captivity.' For the year AD 871, the Annals of Ulster tell how 'Amlaíb and Ímar returned to Áth Cliath from Alba with two hundred ships, bringing away with them in captivity to Ireland a great prey of Angles and Britons and Picts' (the Picts were a people who lived in Scotland and subsequently disappeared, without their language being preserved).

I shall never know for sure how my paternal grandmother's maternal lineage came to Sweden. However, the suspicions aroused in me when I saw the DNA results have dispelled any romantic notions I might once have harboured about the Viking Age.

CHAPTER THIRTY-FOUR

The Mothers

My maternal grandmother, Berta, died of breast cancer several years before I was born. Yet she was a very definite presence throughout my childhood. My mother told me a good deal, and Berta's background was the part of my family history that affected me most. As a child and teenager, I was fascinated by life in the forests of Värmland and the powerful influence of music and poetry.

Berta Gottfriedz, née Turesson, came from Arvika but worked throughout her adult life as an infant school teacher in Tullinge, at that time a small industrial town south of Stockholm, which was dominated by Alfa Laval's separator factory. She was married for a short time, had a daughter, but divorced early on and lived as a single mother in a small flat above the school where she worked. As a child, she had such fiery red hair that she was teased and called names. In later life, her hair was thick, glossy and more coppery than red. At any rate, that is what I've been told; I've only seen black-and-white photos myself. Throughout her life she struggled with surplus weight to some extent, and photos show her in roomy dresses.

'Cheerful' and 'kind' are the words that crop up most often when people describe her, along with the phrase 'she took everyone by storm'. She was always ready to lend others a helping hand. Not long ago I became acquainted with a lady called Dagmar who was 102 when I met her, but still had a crystal-clear memory. As a teenager, Dagmar lived under harsh conditions with unkind relatives in the north of the country. Berta got to hear about this. She intervened, sent money for a train ticket and let Dagmar live with her for a long period of time.

The tears well up in my eyes when Dagmar describes the scene 85 years ago, when my sunny-natured grandmother met her at Stockholm Central Station. What touches me especially are the mentions of constant outbursts of hearty laughter that Dagmar and others come back to when describing my grandmother. My own mother rarely laughed.

Botkyrka Church was full to capacity for my grandmother's funeral. In the address he gave at her graveside, the minister recalled how she was always prepared to offer support to others, be they children in need of help or poor artists.

There wasn't often money for fancy food in my mother's childhood home. Berta preferred to invest in books and original paintings. Several of those paintings are in my possession today.

She was a local councillor in Tullinge for a while. But issues like drains and other practical matters weren't really her style. Her interests ran more to art, literature and poetry. One of her sisters was married to the artist from whom she bought most paintings. Another went out with the singer Ruben Nilsson for a while. And her eldest sister, Olga, was married to the poet Dan Andersson. He died when my grandmother was only 22, but left an impression that lasted for the rest of her life. Dagmar, the 102-year-old, told me how Berta, when she was married and lived in a house, turned a whole room into a Dan Andersson museum.

My grandmother and her sisters were very close. They rang each other every day. And they met in summer at the home of my maternal great-grandmother, Karolina, who lived in Brunskog near Arvika. As a child, my mother spent several summer holidays in Brunskog. Sadly, I could never get her to tell me very much about Karolina.

Going back one generation further to my grandmother's grandmother, Kajsa Gullbrandsdotter, two letters have been preserved. Kajsa wrote them to her granddaughter Berta – my grandmother – when Berta was training to be a teacher in Stockholm. Her handwriting is clear and neat. The spelling and grammar are almost faultless, although Kajsa only attended primary school for a few years. In her letters, she wrote that the winter had been hard, with a great deal of snow. Though Kajsa's husband had managed to reach the village and get to the post office on snowshoes, he had suffered from serious back pain for a long time. However, the two old people had not been completely cut off, as they had had some woodcutters lodging in their cottage for a while.

I find out a little more about my grandmother's grandmother, Kajsa Gullbrandsdotter, in a book by my grandmother's brother, Gunnar. Gunnar Turesson was a ballad singer. One of the poets whose work he performed was his late brother-in-law, Dan Andersson. Gunnar was very popular in the twenties, thirties and early forties. He played the lute, sang and set to music many of the most popular songs of the time, such as '*Jag väntar vid min mila*', '*Flicka från Backafall*' and '*En ballad om franske kungens spelmän*'. But after the Second World War, the Swedes' interest in folk songs cooled off. Gunnar moved back to Värmland and researched folk traditions for a while. He travelled around from village to village with a tape recorder and recorded people singing ancient folk songs and yodelling to call the cows home and keep trolls away. Unfortunately, he wiped the tape clean as soon as he had noted the melodies down, so there are no longer any recordings, they tell me at the Swedish Folksong Archives.

In his autobiography *Visor och skaldeminnen* ('Folksongs and the Memories of a Poet') Gunnar Turesson recounts how he and my grandmother Berta – at the thoughtful Berta's behest, of course – walked several dozen kilometres to help their grandmother Kajsa for a few weeks in the summer of 1917. She had just been widowed, but still had some cows and sheep that she would call home every morning and evening. One day, several years later, Kajsa was so stiff with rheumatism that she was unable to take her birch-bark backpack off in the evening, but had to sleep with it strapped to her back all night. At that point, my grandmother's mother, Karolina, decided to take Kajsa home to Brunskog. She was bedridden for three years before her death. The hearse came to fetch the body, and that was the first time Kajsa ever went anywhere by car.

Gunnar Turesson told of how, as a child, he learned Finnish words and expressions from his grandmother. In the book, he describes how she taught him where 'our Finnish forebears with rye in their mittens had settled after making their way on foot from Savolaks (Savonia) to Gunnarskog'. My mother, too, said we were supposed to be descended from Forest Finns – a group of people who migrated in the sixteenth and seventeenth centuries from Savolaks and the forests of Karelia to Sweden, to practise swiddening, or slash-and-burn farming. 'Your grandmother had a lot of Finnish blood in her veins. That's where you get your high cheekbones from,' I was told on more than one occasion.

But is that really true? At the Gothenburg Book Fair a few years ago, I came across a stand where some genealogy researchers were exhibiting. Perhaps it was there and then that the whole book project started. I asked them to look up my maternal grandmother and great-grandmother. The lightning speed with which they clicked through the register, using my vague information, was impressive. In just a few seconds they managed to find my maternal grandmother Berta, my great-grandmother Karolina, my grandmother's grandmother Kajsa, and her mother, Karin Svensdotter. And that was it. The register had no information about the place of birth of Karin Svensdotter, the mother of my grandmother's grandmother.

But now my curiosity was aroused. A few months later, I consulted the fellow student from my journalism course, the prominent genealogy researcher Håkan Skogsjö. He sat down and took another look. And, in a roundabout way, Håkan managed to discover what the researchers at the book fair had failed to find. He was able to show that Karin Svensdotter was the illegitimate daughter of a woman called Annika Svensdotter, who was born at Hillringsberg Manor, south of Arvika. Sadly, this Annika Svensdotter died at the early age of 45, and by that time she was 'penniless' according to the note in the church records. Annika, in her turn, was the daughter of another woman called Karin, just like me.

One summer day in 2012 I travelled to Hillringsberg to find out more about Karin Gudmundsdotter, the mother of my grandmother's grandmother's grandmother, born in 1735. Hillringsberg lies on Glafsfjorden, an oval lake that derives its name from the beauty of its glittering waters. They glitter as much as ever. The large white manor house has an exceptional setting, with its broad veranda looking out over the water.

In the eighteenth century Hillringsberg was an iron foundry producing mainly malleable iron and nails. There were also mills and saws, and Karin Gudmundsdotter is supposed to have lived in a place called Nedre sågen ('Lower Saw'). This lay next to the waterfall, very close to the manor house and the lake. I wandered about in the area for a while, trying to picture how it would have been in the eighteenth century, when Karin Gudmundsdotter lived here. The booming of the waterfall went right through me, sounding just as it must have back then. The old workers' cottages have burned down and the smithy is gone. It has been replaced by a plant producing solar panels for roofs – one of Scandinavia's largest.

My family history expert, Håkan Skogsjö, thinks it quite possible that my maternal lineage may go back to the Forest Finns who came to Sweden some centuries ago. Glava, as this parish was called, was one of the areas where Finns settled in the seventeenth century. That is why I have arranged to meet K.-G. Lindgren, a local historian and family history researcher who has written several books about Forest Finns. The most recent, called *Där finnar bröt och röjde* ('Where Finns Broke and Cleared the Soil'), explores old crofts and farms in the Arvika area that were once home to Forest Finns.

But K.-G. Lindgren cannot find any evidence at all that Karin Gudmundsdotter might have been the daughter of

Forest Finns. On the contrary, he explains that Swedes and Finns lived totally separate lives: the Finns up in the forests and the Swedes down in the valleys, where there was arable land.

What he has found is that Karin Gudmundsdotter was married to a miller who worked at Hillringsberg. Before that marriage, she had been married to another miller at another industrial settlement, but he died of pneumonia after only one and a half years of marriage, when Karin was 26. Karin's father, Gudmund, from whom she inherited the patronymic 'Gudmundsdotter', was a soldier from nearby Fors farm. He went to war with Karl XII's army in Norway, and after the king's death, Gudmund was appointed parish clerk and organist in Stavnäs, on the other side of Glafsfjorden. That shows he must have been literate and that he must have had a good singing voice. There are only a few kilometres between Stavnäs and Hillringsberg if you take the direct route over the water or the ice. But the usual way to travel was probably to take the ferry over the water a few kilometres further south, where Glafsfjorden is at its narrowest.

On my return from Värmland, I ask Håkan Skogsjö to double-check K.-G. Lindgren's findings. And everything seems to be right. Karin Gudmundsdotter was a miller's wife at Hillringsberg, and she was born at the home of the Stavnäs parish clerk, the daughter of Gudmund and his wife Märta. There is little information about Märta, but she seems to have been born in 1698. Håkan also finds a brief note in church records about the 'mother-in-law' who also lived in the Stavnäs homestead towards the end of her life. She must have been Märta's mother, though she is not mentioned by name.

There is not a word to suggest that any of them might have been Forest Finns. So I go back to my DNA results. I turn to the leader of a Norway-based DNA project for family history researchers with Forest Finns among their ancestors. But the project has no record of anyone with mitochondrial DNA anything like my own rare group,

U5b1b. Some participants belong to the special subgroup that is so common among Sami people, U5b1ba. However, most of the people in the Forest Finns group whose DNA has been tested belong to haplogroups that spread through Europe in connection with farming, such as H, J and T. That tells us something about the origins of the Forest Finns, but it adds nothing to my personal family history.

So there is absolutely no indication that I might be descended in the maternal line from Forest Finns. I suspect that the singer Gunnar Turesson may have embroidered on real life a little in his book, particular as regards the background of his maternal grandmother, Kajsa Gullbrandsdotter. He probably thought it was more exotic and interesting to be a Forest Finn in Värmland. It's clear he was fascinated by Finnish cultural traditions. In his autobiography, he writes vividly of how the Forest Finns in Värmland worshipped the forest god Tapio, and how they nailed bears' skulls to fir trees after the hunt. Gunnar Turesson was very probably right in thinking there were Forest Finns among our Värmland ancestors – but they were not part of the matrilineal ancestry I am investigating now.

***

I am still curious about where Karin Gudmundsdotter's mother Märta, the wife of the Stavnäs parish clerk, came from, and who the 'mother-in-law' was. But Håkan Skogsjö can find no further information. He advises me to contact Peter Olausson, a family historian friend of his who is now a professional historian at Karlstad University, specialising in local Värmland history.

A few days after emailing Peter Olausson, I receive a friendly response. He agrees to look in the archives and see if he can find any more clues about Karin Gudmundsdotter's mother Märta, and her mother, the 'mother-in-law' who lived at Stavnäs.

This was in the autumn of 2012. Silence ensued. Nearly two years later, in June 2014, I received another email from Peter Olausson, which began: 'You've probably abandoned all hope of hearing from me, and justifiably so. There simply hasn't been any time to look for information about your ancestor. I hope you've got some help from someone else.'

But I don't give up that easily. I reply, asking if I can meet Peter in Värmland anyway, even if it's only to ask him some more general questions about life around Glafsfjorden in the eighteenth century. Fortunately, he agrees.

A few weeks later, one oppressively hot day in August, I get off the train at Arvika station. Peter Olausson fetches me by car, and our meeting exceeds all my expectations. He is on leave from his job at Karlstad University, but spends a whole afternoon driving me round to various places around Glafsfjorden that are linked in some way with my family. We visit the beautiful church at Stavnäs, where my forefather Gudmund Erlandsson once worked as clerk, and the place on the nearby shore where his homestead once stood and Karin Gudmundsdotter was born.

By way of a bonus, Peter tells me a good deal about the author Selma Lagerlöf. I didn't realise she had such close links with this area; her childhood home, Mårbacka, was in Östra Ämtervik, several dozen kilometres away. However, she had several relatives around Glafsfjorden. One of them was a pastor in Stavnäs at the same time as my forefather Gudmund Erlandsson was the parish clerk there. Another of her relatives was the pastor of a nearby church, which is now in ruins. He was one of the models for the protagonist of *Gösta Berling's Saga*, Peter tells me. We pass by, and I stand on the spot where the best-known opening in the whole of Swedish literature is set: 'The pastor was mounting the pulpit steps. The bowed heads of the congregation rose – he was there, then, after all, and there would be service that Sunday, though for many Sundays there had been none.' It is true that the pastor was a drunkard, that he was involved in major scandals and that he was dismissed, Peter says. But the rest of Gösta Berling's tale was Selma Lagerlöf's own creation.

We take a look at Hillringsberg Manor, which I visited two years previously, and Fors, the large farm where Gudmund Erlandsson lived as a soldier and met Märta, his future wife. A real major's wife who ruled the roost at Fors was the inspiration for the wife of the major at Ekeby in *Gösta Berling's Saga*, Peter Olausson tells me. It occurs to me that the unusual name Gudmund Erlandsson crops up in another of Selma Lagerlöf's books, *Girl from the Marsh Croft and Other Stories*. An elegant young man by that name is the protagonist and hero of the short story that gives its name to the collection. The story was also turned into a silent film directed by Victor Sjöström. The film had its premiere at Stockholm's 'Röda Kvarn' cinema in 1917, the same year in which my maternal grandmother, Berta, passed her teacher's examination at the teacher training college there. The real-life model for the girl from the marsh croft may have been Gudmund Erlandsson's first wife. My foremother Märta was his second wife, according to church records.

At the linen factory in Klässbol, famous for weaving the tablecloths for the Nobel banquet, we stop for a coffee in an old mill. Peter Olausson has a cup of coffee and a cake, and I buy a bottle of apple juice. The apples were grown and pressed just a few kilometres away. It is so oppressively hot that we choose to sit indoors in the shade.

When we take our seats at the table, Peter suddenly says, 'Oh, there was something else, by the way. I've found the mother-in-law.'

'What?' I reply. After all, it's only a few weeks since he emailed me to say it was impossible to find Märta's mother.

'Yes, but then I did find her after all. Her name was Annika. She was from Stockholm.'

So I've been over to Värmland twice and searched for my ancestors among Forest Finns and other ancient Värmland families, and now the trail leads back to

Stockholm. To the capital, where my parents lived as young medical students, and where I myself have been living for some years now.

'It's quite common for genealogy researchers to suddenly stumble across unexpected coincidences,' says Peter. 'There are times when it seems as if you're going round in circles.'

The document he happened to leaf through before our get-together is a little book called *Anteckningar om Glafva socken i Värmland* ('Notes on the Parish of Glava in Värmland'). It was written by a local historian in the nineteenth century, but the whole edition was bought up and destroyed by a local family who felt they had been misrepresented in some way. It was not until the 1970s that the little book was republished by a genealogy researcher and county librarian. It contains all kinds of notes about people who lived in Glava, and for some reason Peter Olausson noticed the obituary of a church caretaker called Jean Pettersson. Somehow or other – I still don't understand how – he realised that this caretaker was the brother of my grandmother's grandmother's grandmother's grandmother, Märta Pettersdotter.

According to the obituary Jean was born in Stockholm, where his father, Petter Jonsson, worked as a gardener. Some time after his son's birth, the gardener and his wife, Annika Jeansdotter, moved first to Västergötland, then to Jösse in Värmland, and finally to Hillringsberg. This was at the end of the seventeenth century, when Sweden was a great power and it had just become fashionable for the country's wealthy elite to create elaborate gardens in the French Renaissance style. Clearly this also applied to the largest estates in Värmland.

After our coffee, Peter Olausson and I visit the open-air museum of traditional buildings in Glava, where we look at some more documents that confirm Peter's findings. Annika Jeansdotter was the wife of a gardener at Hillringsberg, who had moved from Stockholm with her husband. Her daughter Märta grew up there and was later employed as a maid at nearby Fors farm. It was there she met Gudmund Erlandsson, the soldier who was later appointed parish clerk at Stavnäs, and they moved a few hundred metres away to the other side of Glafsfjorden. A generation later, Märta's daughter, Karin Gudmundsdotter, moved back over Glafsfjorden to Hillringsberg and became the wife of a miller there. For five subsequent generations, women moved no more than a few kilometres within the area that is now the municipality of Arvika. When my grandmother Berta started studying at the teacher training college in Stockholm during the First World War, she returned to the city her foremother had left about 240 years previously.

But where did Annika Jeansdotter, the gardener's wife, come from originally? She is very likely to have been a relative newcomer to Stockholm. When Sweden was a great power, the rapidly expanding capital was a magnet, attracting large numbers of people from other places – just as it does today. I am one of those who have moved to Stockholm for work reasons.

Was Annika Jeansdotter originally from Värmland after all? Written sources are probably never going to take me any further back. What remains are the DNA results. One day I receive an email that ends with the words: 'Best wishes from your very distant relative Tomas.' The man who wrote the letter is the nearest match to me so far on that branch of my family tree. Our mitochondrial DNA sets differ only on two points. His earliest female forebear also lived in Värmland in the eighteenth century, but in the parish of Norra Råda, which lies further north, in the municipality of Hagfors. She was also called Annika, and her daughter's name was Karin.

***

My own mother was not called Annika, but Anita. She, too, lived in Stockholm for a few years while she was studying medicine at the Karolinska Institute. There is a black-and-white photograph taken in the flat where she and my father lived as newlyweds.

The little one-roomed flat is full of paintings and beautiful furniture, most of it inherited from my grandmother, who had died a few years previously. She spent much of her income on paintings by artist friends and on other beautiful objects. The photo shows my mother playing an old square piano, an old-fashioned form of the instrument. Above the piano hang a violin and a guitar. There is a writing desk in masur birchwood and a substantial table of solid pine. All the furnishings were chosen with the help of relatives in Värmland – my grandmother's musician brother, Gunnar, and her father, who built organs for a living.

The photograph shows my mother as a slim young woman with fashionably styled hair. She is recently married and a student on one of the country's most prestigious university courses. The picture must have been taken around 1957, which was a great time for our maternal line. Only 150 years had passed since my grandmother's grandmother's grandmother in the maternal line, Annika Svensdotter, died at 45 as the penniless widow of a blacksmith. During these 150 years, Swedish society underwent unprecedented development from poverty to prosperity. The same was true of my family.

About 10 years after the photo was taken, the problems began to pile up for my mother as well. Illness and other problems destroyed our family. The riches of her genetic make-up were beyond money; she was very gifted and full of energy. But there was a dark side to that heritage. Humanity pays for artistic gifts and creativity, and the toll is mental illness.

That is the way it has been throughout European history, ever since the Ice Age people created their cave paintings and carved figures from ivory. And I firmly believe that that dual heritage was in our baggage a long time before some of us left Africa to people Europe and the rest of the world.

I have been lucky in the genetic lottery. I have been spared serious mental illness, but have nonetheless inherited a generous portion of creativity and energy. That's a winning ticket to be grateful for. Rather than fret over the impact that

illness and other problems had on my childhood, I can be
thankful for the strength and the abilities I was born
with. The different sides of my biological heritage are inter-
connected. The bad comes with the good, the good with
the bad.

That's the nature of genetics. You can't say that certain
genetic variants are 'better' than others. Inherited traits can
be advantageous for some individuals in particular circumstances but represent a drawback in others. The British
author Richard Dawkins made a big splash in the 1970s with
his popular science book *The Selfish Gene*. He has subsequently
tried to alter the concept somewhat and tone it down. But
the damage has already been done. The punchy title gave a
whole generation a simplified, indeed erroneous idea of how
DNA works. This prolonged the political polarisation of
the 1930s and 1940s, when Nazis were keen to stress blood,
earth and heredity, while Joseph Stalin and his followers
in the Soviet Union regarded heredity as 'bourgeois' and
'counter-revolutionary'.

Genes are not selfish, nor do they exist in isolation. They
work together in large, complex constellations of DNA.
DNA is passed on to each new generation, not in the form
of individual genes, but in large bundles.

Rather than 'selfish genes', I would prefer to use the term
'two-faced genes'. Genes can be good or bad, depending on
the environment in which they occur. They can result in
mental illness or great creativity; in a well-filled physique,
which enables the individual to survive in harsh surroundings,
but also causes overweight if there are ample amounts of food
available. Hypersensitivity to sensory impressions helps the
hunter to locate his quarry, but can be disastrous in a classroom
or an office. What is good or bad depends on the combination
and the context.

And inherited traits are just the beginning. We are also
formed by all our experiences, from the time we are in the
womb, and throughout our lives. Our DNA is also affected to
some extent by the experiences of previous generations

through mechanisms known as epigenetics, which scientists are only just beginning to understand.

Together, nature and nurture condition our identity and our health. They belong together. Only the very ill-informed now believe there is a contradiction between these two poles; the very ill-informed, and those who are blinded by ideology.

CHAPTER THIRTY-FIVE

The Legacy of Hitler and Stalin

Though DNA research is advancing in huge strides, it is dogged by dark forces reminiscent of the totalitarian ideologies of the 1940s.

The voices that hark back to the Stalinist era are something I have been aware of for a long time. Such tones were commoner 20 years ago, when I started writing about the achievements of the new biotechnology. At that time, I quite often got readers' letters voicing sweeping prejudices against genetics. Fellow journalists would sometimes demonstrate a total lack of nuance, coming out with forms of words that might have been taken straight from Stalin's witch trials of biologists. I will never forget the time a cultural commentator who was prominent back then suddenly burst out – and this is a direct quote – that 'all geneticists are fascists of a kind'.

A visit to Russia's Vavilov Institute made me even more sensitive to that sort of knee-jerk criticism of DNA research. The Institute's premises, still a feature of central St Petersburg, housed the world's first major seed bank. When stem rust threatens wheat harvests worldwide, when the global climate heats up and drought becomes more widespread, plant breeders will be able to search seed banks like this for resistant genetic material to grow the crops of the future.

The Institute is named after Nikolai Vavilov, one of the world's foremost plant geneticists in the first decades of the twentieth century. He travelled on five continents, collecting seeds from wild species and traditional varieties from all the farming environments imaginable. One of the motives that drove him was the desire to end hunger and famine in Russia and the rest of the world. He sought to improve and secure food resources by collecting the raw material for new and better crops. But he was also passionate about fundamental research. He wanted to know where agriculture was first developed. I have found myself thinking of Vavilov a good deal while working on the parts of this book that deal with the genesis of agriculture. If only he had had access to the new DNA studies on wheat, beans and other crops that have enabled today's geneticists to pinpoint the birthplace of early agriculture as the border regions of Turkey and Syria.

From 1924 Vavilov was the director of the seed bank in Leningrad, as the city was then called, and he also became the head of the genetics department of the Soviet Union's Academy of Sciences. However, in the early 1930s a young agricultural engineer from the Ukraine called Trofim Lysenko began to scheme against him. Lysenko claimed that the importance of genetic traits was overrated. Mendel's laws of heredity were mistaken, and environmental influences could, in fact, be inherited. Wheat and other crops would be able to adapt to the harsh climate of Siberia if only they were treated in the right way to withstand the cold.

Lysenko got the ear of Joseph Stalin, as his rhetoric was an excellent match for the prevailing Soviet jargon. Just as the new Soviet man and woman would develop under new, more favourable conditions, so crops would grow stronger, healthier and better in a socialist society. Genes were 'bourgeois' and 'counter-revolutionary'; environment was all. The working conditions of serious Soviet geneticists continued to deteriorate. Many of them were imprisoned in the 1930s. Vavilov was one of those who kept going longest, although he was clear in his criticism of Lysenko. But in 1940 he was arrested by Stalin's henchmen. He was imprisoned and sentenced to death. The sentence was subsequently commuted to 20 years' imprisonment, but Russia's greatest plant geneticist died of starvation in a gulag in 1943.

Leningrad was under siege by the Germans at that time, and was cut off for 900 days. At least a million inhabitants died, largely through starvation − nearly half the city's people. The staff of the seed bank guarded the collections with their lives. They could have made porridge from the oats and pea soup from the stores of dried peas. Yet they did not do so. On my visit to the Vavilov Institute, I saw photographs of 15 members of staff who died at their posts during the siege.

Lysenko's erroneous doctrines damaged biological and genetic research in the Soviet Union, China and eastern Europe for many decades after that.

That is what springs to mind when I hear lazy, sweeping criticisms of genetic research. However, there is another side to this. There are DNA tests on sale in Hungary which, so it is claimed, can establish that the person tested has no Jewish or Roma forebears. A science journalist who used to be well respected, Nicholas Wade of the *New York Times*, with a background at both *Science* and *Nature*, recently published a book that makes a series of problematic assertions. These include his thesis that natural selection has resulted in differences in IQ, educational outcomes, political systems and economic development between different parts of the world. Over 100 of the world's foremost DNA researchers – including Svante Pääbo and Mattias Jakobsson from Sweden, Eske Willerslev from Denmark and the American David Reich – have signed a sharply worded petition in which they distance themselves completely from Wade's theses. They write: 'We reject Wade's implication that our findings substantiate his guesswork. They do not.'

These events hit me hard. Wade was a fellow journalist whom I once respected; he has worked on some of the world's leading science desks at the world's most highly respected journals. He has had access to the same published research on genes that I have been following over the last 20 years. And yet he has gone off on the wrong track. Despite everything, he has started to make claims about conclusions that scientists such as Pääbo, Jakobsson, Willerslev and Reich have certainly never drawn.

I have every confidence in the leading Swedish exponents of genealogical research. They show intelligence and discernment in avoiding the potential pitfalls of both genetics and ethics. Their shared websites contain rebuttals of commentators who draw any far-fetched or erroneous conclusions. They hone the wording they themselves use, so as to avoid any overinterpretation or misunderstanding of their material.

Unfortunately, there are other blogs and discussion groups that are less scrupulous. Some of the most deplorable expressions I stumbled across came up when I was trawling the net for information about my paternal grandfather's haplogroup, R1a. There are a number of people out there in cyberspace who are trying to spread the notion that R1a is an 'Aryan' group and that those belonging to it have superior characteristics. Such voices are widely heard in India, but they occur in Europe too.

That is a very regrettable state of affairs – but it is not a reason to abandon DNA as a new and important tool for researching the origins of humankind. 'We can't let Hitler dictate what subjects we can research, 50 years on.' That was Svante Pääbo's riposte to the Senate of the Max Planck Society when they were discussing whether it would really be acceptable for Germany to set up a new institute for anthropological research, given the role the old anthropological institute had played in the Holocaust.

Instead, we must learn from history, says Pääbo. One important lesson is that science must be based on facts. It is essential that scientists work empirically on the basis of observations and experiments, not just theories; that lessens the risk of their being seduced by their own prejudices. That racist theories crop up is hardly surprising, given the human penchant categorising each other. But the new DNA research provides no grounds for such beliefs.

Pääbo refers to certain huge projects involving comparisons between the DNA sets of thousands of people from different countries and continents. These have cost a great deal of money, one estimate being US$120 million (£99 million).

'What have we found out for all the dollars spent? Well, there are some local differences in skin pigmentation and variations in immune defence and our capacity to break down what we consume, such as lactose and alcohol. Thanks a lot – I already knew that,' says Pääbo. On the other hand, he continues: 'But maybe it's been worth spending all that money just because of what we *haven't* found. When it comes to the

way brain cells work, for instance, there is *no* difference to
be found between different groups and countries. We know
that today.'

Another of the researchers who has contributed most
to the findings on which this book is based is Mattias
Jakobsson. On my way to his office in the Centre for
Evolutionary Biology on the periphery of central Uppsala, I
walk past the *Dekanhuset*, which housed the State Institute
for Racial Biology in the 1920s and 1930s. The building,
which stands next to the cathedral and opposite the
archbishop's residence, is owned by the National Property
Board of Sweden. There is a small signpost in front of it
with some information about the various activities that once
took place there. For the years between 1869 and 1951, it
states briskly that the premises were used by organisations
including the City of Uppsala, Uppsala University and a
primary teacher training college. There is no mention of
any Institute for Racial Biology. Precious little seems to
have been learned from history.

Mattias Jakobsson's office is just a few hundred metres from
the location of the world's first Institute for Racial Biology.
He has reflected a great deal on what he, as a geneticist
working in Uppsala, can contribute with regard to that
institute and the history of Uppsala University. He already
holds regular popular science talks on genetic variation
among people in different parts of the world. In the medium
term, he plans to work together with experts in the history
of ideas to set up a special course on ethics and history for
students of genetics. Jakobsson shares Pääbo's view that
scientists today should not allow their practice to be dictated
by the actions of Hitler and the 'racial biologists'.

The Sami people, for example, were subjected to Herman
Lundborg's quasi-scientific programme of cranial measurement. They were humiliated by being photographed naked
and by offensive statements about race and innate disposition.
'But studying the origins of Sami people today shouldn't be
any different from studying the origins of other Swedish
people. It would be worse if we chose to avoid the issue by *not*

studying particular groups. Hopefully this will be a less sensitive issue in the future,' says Jakobsson.

***

In my view, private individuals should not avoid using DNA to investigate their own family history either, just because there are instances of abuse and misinterpretation. However, there are ethical problems one needs to be aware of.

The most obvious risk, as I see it, is that of unexpected and unwanted information about family relationships. I am thinking in particular here of cases where a person turns out to have a different father. Cases of mistaken paternity are not as common as is sometimes asserted; I have heard figures of up to 10 per cent, but that would hardly apply under normal circumstances. However, there are many children who have grown up with the wrong information about their father (or mother, though that happens less often). In such cases, the truth can come as a shock, not just to the person who has chosen to be tested, but to other family members as well. You need to be psychologically prepared for such situations before having a DNA test carried out. Moreover, many people looking into their family history persuade relatives to be tested, and it is even more important in such cases to think about the possibility that unexpected information might turn up.

Another theoretical issue is the possibility of information about hereditary conditions getting out. The first DNA test I ordered – carried out by the Icelandic firm deCODE – included information of that nature, though it was fairly general. deCODE has since gone bankrupt. 23andMe used to provide information both about the risk of various medical conditions and on family relationships, but the US Food and Drug Administration has since imposed limits on the medical side of their activities.

The genetic tests for family history researchers mentioned in this book are not designed to provide medical information.

My personal view is that historical and medical research should be kept separate as far as possible. Tests for serious hereditary conditions should be dealt with through the professional healthcare system, not by private individuals buying tests online. But what complicates matters is that a person with the right expertise could work out a good deal of information about the risk of various medical conditions by analysing the basic data included in more detailed family history tests. So there are good reasons to be careful with your password and about confidentiality in general, and to think twice before passing on any information to others.

A third risk is that DNA data may leak out into the public domain through human curiosity. Personally, I can't see why an outsider would want any information about my DNA. I don't think I'm that interesting. But people have different needs as regards privacy. Many chat away cheerfully about their private lives and thoughts on social media, while others prefer to be more circumspect.

There may be situations where a DNA sequence could have gossip value. If the haplogroups of the King of Sweden were made public, the newspapers would definitely write about the subject. But the articles would probably be short. The news would blow over in a day or so. Personally, I can only see advantages. If the Bernadotte family's Y-chromosomal DNA were public, it would put paid to shady businesses trying to peddle information about whether particular individuals are related to the king.

It is easier to foresee potential threats to groups that have been particularly vulnerable in the course of history, such as the Jews, Roma, Afro-Americans and Sami. But the fact is that DNA research into family history is often even more popular within ethnic groups that share a strong identity. There is a lively interest in the history of the group, and DNA tests can sometimes contribute information that cannot be supplied in any other way.

All DNA researchers – both the professionals and amateur researchers of family history – need to reflect on the possible ways in which their results may be misused. Professional researchers have ethics committees that lay down codes of practice governing their work. But we amateurs also have ethical responsibilities we cannot abdicate.

In my view, turning aside from genetic research – in the spirit of Trofim Lysenko – is a very poor solution. The important thing is to inform ourselves, and to try and understand what our DNA sequences can tell us.

The Tree and the Spring

Half a year after my mother's funeral, we hold the urn placement ceremony in Botkyrka Church. It is a relaxed, low-key occasion by comparison with the funeral in Gothenburg.

It was early summer then, the lilacs were in bloom and old friends appeared unexpectedly. We sang hymns about the blossom time, about the ages to come and the generations to follow. It was a solemn, but also slightly stilted ceremony.

Only my brother and myself, with our respective partners, attend the ceremony in Botkyrka. The grandchildren are allowed to stay in school. The November weather is at its worst. Slushy sleet covers the ground. A hard wind blows out the matches when I attempt to light a candle.

Botkyrka Church is where my mother was confirmed. She told me how she used to ski over the fields from Tullinge to get to classes. Having only known her as an adult, I am surprised she could ski such a long distance; it must have been almost 10 kilometres (6¼ miles). Today, the church and churchyard are squeezed in between a broad motorway and the heavily built-up suburb of Hallunda, with its blocks of flats from the housing programme of the sixties and seventies. Nearly 80 per cent of people living in Hallunda are immigrants, and I have to ask several people at the metro station before someone can tell me the way to the church.

Yet the church has been there since 1129, the time when a local peasant lad named Botvid is said to have introduced Christianity to the area after a journey to England. The story goes that Botvid was killed by a thrall whom, in accordance with his new convictions, he was just about to set free. At the place where the pall-bearers put down Botvid's coffin, a spring burst forth, and when sick people touched the coffin they were made whole, according to the legend.

Fifty years later, the wooden church built in 1129 was replaced by a stone church. Parts of it remain, and that is where we begin the ceremony. We sit for a long while contemplating the urn, which is placed on a small table before the altar. A kindly caretaker plays a mix of classical music through the loudspeakers. Finding some of the music rather too light and bland, I request Bach instead.

Botkyrka Church is a few kilometres from the spring named after the same saint. But Sweden's first churches were often built on ancient cult sites from pre-Christian times, at sacred springs and near sacred trees. People simply continued to visit the same sacred groves as in earlier times; they merely changed the content of their rites. In the new churches they worshipped God, Christ and the Virgin Mary, whereas at the ancient cult sites they had prayed and made offerings to other gods and supernatural beings.

When, together with the historian Peter Olausson, I visited the outdoor museum in Glava, the Värmland parish where my foremothers lived for many generations, we viewed an old drawing made when the first church at Glava was demolished. That church had been built in the twelfth century, at about the same time as Botkyrka Church. 'Tree stump' was written on the drawing, in front of the altar. The fact that the old church in Glava was built over a tree stump was news even to Peter Olausson, an expert in local history. But he told me about other examples of early churches in Sweden where the altar was placed on top of old trees that had been felled. The stump was built into the new shrine as a precaution. Altars were also sometimes built over an ancient spring. The new religion was superimposed on the old one.

It is common to depict your ancestry in the form of a family tree. I have simplified matters still further in this book by focusing on three straight lines of descent. My mitochondrial DNA shows I am descended from a woman who came to Europe during the Ice Age, about 40,000 years ago, a woman we can choose to call Ursula. I am also descended from a woman we can call Helena, one of the group of people that brought farming to Europe 10,000 years ago. And in all probability my forefathers, whom, for simplicity's sake, we can call Ragnar's descendants, came to Jutland together with a new, expansive pastoral culture, and later to Sweden in connection with the Bronze Age.

But these three lines are only a very limited part of my ancestry. I also had a maternal grandfather, for instance, although he is not included in this book. For one thing, I did not have the same emotional bonds with him, and for another I have not had the same opportunity to investigate his Y-chromosomal DNA. Yet he, too, is part of my family. The family grave in Botkyrka, where my mother's ashes now rest, belongs to his mother's side of the family.

Going back two generations, I have four ancestors: my maternal grandparents on the one side and my paternal grandparents on the other. Each had two parents, so if I go back a further generation I can count eight ancestors. After that, the numbers are 16, 32, 64,

128, 256, and so on. In many cases, my forebears had children together, which complicates the mathematics. Which individuals, out of all these forefathers and foremothers, passed on their genes and their traits to me is a matter of chance.

That is why the tree gives an incomplete picture of our history. It can only show a single individual's ancestry over a few generations; after that, there are not enough branches and roots on the tree.

As for the history of the population as a whole, the tree image is also too simple and inaccurate. People have migrated in all sorts of directions in the course of history. Europe, for example, received at least three major waves of immigration in prehistoric times, as well as many other smaller ones. The migrants came on foot from the south and the east, they paddled along rivers and the coastline, and they travelled by ox-drawn wagon and on horseback.

The genealogical tree also gives an incomplete picture when it comes to a biologist's description of evolution. Species do not evolve as directly or as neatly as the branches of a tree would suggest. There are floods of genes that complicate the picture – such as when anatomically modern people in the Middle East had children with Neanderthals 54,000 years ago. This injected a small lateral dose of DNA and genes into the genetic material of modern humans, and the traces can be seen in our DNA to this day.

To describe our origins in a biologically accurate fashion, the image of the tree needs to be complemented by that of the spring. Water flows out of the spring, an amorphous stream running in different directions. Our DNA and our genes disperse in an analogous way. One of 'Ursula's daughters' was my foremother, belonging to haplogroup U5b1b1, the woman who lived some 15,000 years ago at the end of the Ice Age, probably in part of Spain. Some of this woman's modern-day descendants live among the Sami of northern Scandinavia, while others are to be found among the Berbers of the Atlas Mountains, in North Africa.

Everyone in the world is descended from a woman we can call Eve. This woman lived in Africa about 200,000 years ago. As descendants of Eve, we bear largely identical DNA. We are differentiated only by tiny variations. These tiny variations tell the story of how our forebears once went forth to people the earth. The mutations show how we migrated in all directions from Africa. My family – like everyone's – can be compared to a great tree with many branches. But we are like water too. Our heredity both diverges and converges – as in the wellspring of life and humanity.

Questions and Answers about DNA

Which company and product should I choose for a DNA test?

That depends on what you want to know and how much you are prepared to pay. Firms come and go, and their product ranges change. Two commercial firms that have specialised in genealogy research and individual historical origins are Family Tree DNA and 23andMe. National Geographic sells DNA tests focusing on individual historical origins.

There are many reliable websites that provide up-to-date information for people interested in buying their first test. One of them is run by SSGG, a non-commercial network of Swedish genealogy researchers working with DNA, of which I am a member. Its international counterpart is the International Society of Genetic Genealogy, ISOGG.

How does a test for a private individual work?

Usually, you'll order a test kit online from one of the firms mentioned above. Follow the accompanying written instructions to collect a small DNA sample; this may involve taking a swab from the inside of your cheek using a plastic spatula, or spitting into a test tube. Then send your sample off to the test firm. A few weeks later, you will receive a response online, which will contain some basic information. It may often be a good idea to approach experts who can provide further details, such as project administrators linked with the firm but working on a non-profit basis, or – if the test is particularly sophisticated – other firms that are specialised in interpreting DNA data. One such is YFull, based in Russia.

Who should I have tested first?

That depends on what you want to know. But one practical principle when using DNA as a tool for researching your family history is to have the oldest individual in a particular line tested before it's too late.

What is DNA?

The acronym stands for deoxyribonucleic acid, a chemical present in nearly all our cells that passes on our biological traits. The DNA molecule comprises four units known as nucleobases. These are called adenine, guanine, cytosine and thymine, and are normally shortened to A, G, C and T. One individual's genetic material comprises over three billion As, Gs, Cs and Ts. Just a few nucleobases in every thousand differ between one randomly selected individual and another, the rest being identical. Although the variations are so small, they can provide a great deal of information about an individual's origins and genealogical relationships.

What are genes?

Genes are specific components of our DNA that code for hereditary characteristics. They account for no more than a tiny percentage of the whole DNA molecule. The rest of our DNA serves a regulatory purpose, has an unknown function, or is simply 'in-between DNA'. It is more important to keep sequences from within the genes confidential, as they can reveal information about the risk of particular medical conditions and other sensitive matters. It is advisable not to let out such information in public. However, it may be a good idea to let project administrators and other experts you may choose to consult know the whole sequence, including the parts that code for information. This may enable you to obtain better information about genealogical relationships.

What are mitochondria?

Mitochondria are tiny structures within cells but outside the cell nucleus that contain small amounts of their own DNA. We inherit our mitochondria solely from our mothers. It is simpler and cheaper to test mitochondrial DNA than other DNA, as we have large numbers of mitochondria in every cell. The downside, however, is that they contain a very small proportion of our total DNA – only about 16,000 nucleobases, compared with the three billion nucleobases in the cell nucleus. One of their advantages is that mitochondria are passed on unchanged from mother to child for many generations. This means they can be used to trace a person's maternal lineage a long way back in time.

What is nuclear DNA?

Nuclear DNA, which accounts for a large proportion of all DNA, is found in the cell nucleus. Analysing nuclear DNA is generally more difficult and costly than analysing only mitochondrial DNA. The more detailed the analyses you order, the more expensive they tend to be.

What are Y chromosomes?

Y chromosomes are a part of nuclear DNA passed on only by men to their sons. They are sex chromosomes that normally give a man his sexual traits. By comparing mutations in Y chromosomes, a man's paternal lineage can be traced a long way back in time. The most detailed Y-chromosome tests provide far more information than the most detailed tests on mitochondrial DNA, so they will improve your chances of being able to make links with historical sources and the present day.

What are X chromosomes?

X chromosomes are also part of our nuclear DNA. Both men and women inherit an X chromosome from their mother. Women also inherit an X chromosome from their father. This is why women normally have two X chromosomes, while men have one X chromosome and one Y chromosome. The particular way in which X chromosomes are passed on mean they can be used to establish genealogical relationships by means of autosomal tests (see below).

What is autosomal DNA?

Autosomal DNA is DNA from the cell nucleus that an individual inherits from both their father and their mother, and which is randomly mixed each generation. The most common and inexpensive commercial tests for people looking into their family history are based on autosomal DNA. They can provide information about siblings, parents, cousins, second cousins and other relatives at up to seven degrees of kinship. However, this naturally works only if the other people concerned have also been tested.

The strict biological definition of autosomal DNA is DNA that is neither mitochondrial nor from sex chromosomes. However, genealogy companies also include X chromosomes in autosomal DNA tests, making it easier to trace your relatives.

What is a haplogroup?

A haplogroup is a group of individuals with the same set of mutations, who thus share a common female or male ancestor some considerable way back in history. Another way of expressing the concept is that all the individuals in a haplogroup belong to the same branch of a genealogical tree.

What does STR mean?

STR stands for 'short tandem repeats'. A short tandem repeat is the location where the DNA molecule gets stuck, as it were, and a short sequence is repeated a number of times. The number of repetitions at each location is hereditary and differs between individuals. This makes STR a practical tool for ascertaining whether a family relationship exists. It has been used in forensics since the 1980s and is now also a popular tool in genealogical research, especially for comparing Y chromosomes. There is no known link between STR and hereditary traits, which makes its use less controversial. The commercial Y-chromosome tests known as Y-DNA37, Y-DNA67 and Y-DNA111 calculate the number of repeats at 37, 67 and 111 locations on the Y chromosome respectively. The higher the number, the greater the precision and the closer the genealogy researcher can get to the present day. A compromise that often provides good value for money is to test a moderate number of STRs and combine the results with tests on single nucleotide polymorphisms, SNPs.

What does SNP mean?

SNP stands for 'single nucleotide polymorphism'. It is a single location on the DNA molecule, the nucleobase, where people can have different variants. SNPs are often called 'snips'. Geneticists often test a selection of snips instead of all of a person's hereditary material, as this is generally simpler and less costly. Snips have long been used in medical research, but they are now also beginning to be used in DNA-based genealogical research. There are thousands of different kinds of SNP tests, so you should discuss your needs with a specialist before selecting one, to be sure of which test will provide the most relevant information in your particular case.

What are HVR1 and HVR2?

These abbreviations stand for 'hypervariable region' 1 and 2. They are two small parts of our mitochondria where mutations occur more frequently than in the rest of our DNA. Analysing them shows which mitochondrial haplogroup a person belongs to. But this is a rough division which can only provide information about a person's origin in the maternal line several thousand years ago. Looking into closer genealogical and family relationships and getting closer to the present day requires more complete DNA analyses of whole mitochondria.

What are CRS and rCRS?

CRS stands for Cambridge Reference Sequence, which is used for the purpose of comparing mitochondrial DNA. This involves identifying the mutations that differentiate a mitochondrion from the first human mitochondrion ever sequenced; this was done in Cambridge in the 1970s. The improved version is known as rCRS, the revised Cambridge Reference Sequence.

What is RSRS?

RSRS stands for Reconstructed Sapiens Reference Sequence. It is a more recent method for comparing mitochondrial DNA that has been in use since 2012. This method is based on the sequence that, according to scientists' calculations, was borne by the female ancestor we all share, 'mitochondrial Eve'.

What are mutations?

Mutations are random changes in the DNA molecule whereby certain units — nucleobases — are replaced, added or disappear.

There are roughly 30 new mutations per human generation. The older a father is, the more new mutations his children will have. Mitochondria undergo one mutation every 2,000 years on average within the same line of descent.

What do all the letters and figures in my mitochondrial DNA result mean?

They are terms for different types of mutations. To give an example, C40624T means that the nucleobase called cytosine, C, has been replaced by thymine, T, at location 40624. If the result begins with A, T or G instead, and ends with a different letter, this means that another nucleobase – adenine, thymine or guanine – in the reference sequence used (see rCRS and RSRS above) has been replaced.

Sometimes the result is followed by an exclamation mark; for example, C40624T! The exclamation mark stands for a reverse mutation. This means that the original version – in mitochondrial Eve or the Cambridge Reference Sequence – had a T at exactly the same location. Within the haplogroup as a whole – the larger branch to which the sample belongs – that T has been replaced by another nucleobase. But the side branch on the tree to which the tested sample belongs has since mutated back to the original T.

Sometimes you can see 315,1 or 315+C. That means that an extra nucleobase, a C in this instance, has been added after position 315. The opposite of that would be 315D. D stands for deletion, meaning that the nucleobase at the location indicated has been deleted.

There are a number of other abbreviations for specific types of mutations.

References, Further Reading and Travel Tips

The Troll Child: 54,000 Years Ago

The chapter entitled 'The Troll Child' is the product of my own imagination. However, it is based as far as possible on genuine research data concerning our contacts with Neanderthals. References to this research are provided in Chapters 2 and 4.

I decided early on, at the recommendation of the Israeli archaeologist Ofer Bar-Yosef, to set the scene in Galilee. When I started writing, there was no definite proof that anatomically modern people and Neanderthals had coexisted in Galilee during the period in question. It was not until January 2015, when my manuscript was already finished, that such proof was published. It took the form of a description by Israeli archaeologists of the cranium of an anatomically modern human found in Manot cave and dated at about 55,000 years.

As far as researchers can judge from the shape of the Manot cranium, the individual in question was not the result of hybridisation between Neanderthals and modern humans. This suggests that interbreeding took place somewhat later – one of the reasons for my choice of the subtitle 'the first 54,000 years'.

The botanist Mariette Manktelow – who is, among other things, an expert on biblical flora – explained to me what the local vegetation may have looked like at different times of the year. Why do I think it was the Troll Child's father who was a Neanderthal, not his mother? Though that is speculative, there are DNA findings that support such guesswork. No one alive today has been found to bear mitochondria from Neanderthal ancestors. In other words, there seems to be no unbroken matrilineal link between the Neanderthals and us. Moreover, there is less genetic material from Neanderthals in our X chromosomes than in the rest of our DNA. One conceivable explanation for this is that 'troll children' may have had Neanderthal fathers, while their mothers were modern humans.

Bar-Yosef Mayer, D., Vandermeersch, B., & Bar-Yosef, O. (2009). Shells and ochre in Middle Paleolithic Qafzeh Cave, Israel: indications for modern behavior. *Journal of Human Evolution, 56* (3), 307–314. doi:10.1016/j.jhevol.2008.10.005

Baruch, U. (1986). The late Holocene vegetational history of Lake Kinneret (Sea of Galilee), Israel. *Paléorient, 12* (2), 37.

Fagan, B. M. & Durrani, N. (2013). *People of the Earth: an introduction to world prehistory* (14th edition). Boston: Pearson.

Hershkovitz, I., Marder, O., Ayalon, A., Bar-Matthews, M., Yasur, G., Boaretto, E., et al. (2015). Levantine cranium from Manot Cave (Israel) foreshadows the first European modern humans. *Nature*, doi:10.1038/nature14134

Bar-Yosef, Ofer. Email, July 2013.

Manktelow, Mariette. Email, July 2013.

Neanderthals in Leipzig

Leipzig is a pleasant city. I would recommend readers who may wish to visit other places described in this book, such as the Museum of Prehistory in Halle, the Goseck solar observatory and the Nebra Ark (Arche Nebra), to use it as a base.

Take the opportunity to dine at Bayerischer Bahnhof, the former railway station turned brewery and beer garden. If you can cope with rather inconsistent standards and a steep staircase, Antikhotel Völkerschlachtdenkmal is an experience; affordable, it offers excellent breakfasts, cycles for hire and fine antique furnishings. However, it is some way out of the city centre.

While the Max Planck Institute for Evolutionary Anthropology is not generally open to the public, its Zen-inspired architecture can be viewed from the exterior and the reception area.

For a deeper understanding of Svante Pääbo's work, I would recommend reading his book *Neanderthal Man*, which touches on his own life and gives a great deal of information about his research.

Abi-Rached, L., Jobin, M., Kulkarni, S., McWhinnie, A., Dalva, K., Gragert, L., et al. (2011). The shaping of modern human immune systems by multiregional admixture with archaic humans. *Science, 334* (6052), 89–94. doi:10.1126/science.1209202

Auel, J. M. (2010). *The Clan of the Cave Bear*. London: Hodder & Stoughton.

Bojs, K. & Bratt, A. (2011). *Vikten av gener: hur DNA påverkar din vikt* [The Weight of Genes: how DNA affects your weight] (1st edition). Stockholm: Natur & Kultur.

Cann, R., Stoneking, M., & Wilson, A. (1987). Mitochondrial DNA and human evolution. *Nature*, *325* (6099), 31–36.

Green, R. E., Krause, J., Ptak, S. E., Briggs, A. W., Ronan, M. T., Simons, J. F., Du, L., Egholm, M., Rothberg, J. M., Paunovic, M., & Pääbo, S. (2006). Analysis of one million base pairs of Neanderthal DNA. *Nature*, *444*, 330–336.

Green, R., Krause, J., Briggs, A., Maricic, T., Stenzel, U., Kircher, M., et al. (2010). A draft sequence of the Neandertal genome. *Science*, *328* (5979), 710–722. doi:10.1126/science.1188021

Hershkovitz, I., Marder, O., Ayalon, A., Bar-Matthews, M., Yasur, G., Boaretto, E., et al. (2015). Levantine cranium from Manot Cave (Israel) foreshadows the first European modern humans. *Nature*, doi:10.1038/nature14134

Joordens, J. A., d'Errico, F., Wesselingh, F. P., Munro, S., de Vos, J., Wallinga, J., et al. (2014). *Homo erectus* at Trinil on Java used shells for tool production and engraving. *Nature*, doi:10.1038/nature13962

Krings, M., Stone, A., Schmitz, R., Krainitzki, H., Stoneking, M., et al. (1997). Neandertal DNA sequences and the origin of modern humans. *Cell*, *90* (1), 19–30.

Noonan, J. P., Coop, G., Kudaravalli, S., Smith, D., Krause, J., Alessi, J., Chen, F., Platt, D., Pääbo, S., Pritchard, J. K., & Rubin, E. M. (2006). Sequencing and analysis of Neanderthal genomic DNA. *Science*, *314*, 1113–1118.

Prüfer, K., Racimo, F., Patterson, N., Jay, F., Sankararaman, S., Sawyer, S., et al. (2014). The complete genome sequence of a Neanderthal from the Altai Mountains. *Nature*, *505* (7481), 43–49. doi:10.1038/nature12886

Pääbo, S. (1984). Über den Nachweis von DNA in altägyptischen Mumien. *Das Altertum*, *30*, 213–218.

Pääbo, S. (1985). Molecular cloning of Ancient Egyptian mummy DNA. *Nature*, *314* (6012), 644–645.

Pääbo, S. (1995). The Y chromosome and the origin of all of us (men). *Science*, *268* (5214), 1141–1142.

Rodríguez-Vidal, J., d'Errico, F., Giles Pacheco, F., Blasco, R., Rosell, J., Jennings, R. P., et al. (2014). A rock engraving made by Neanderthals in Gibraltar. *Proceedings of the National Academy of Sciences of the United States of America*, *111* (37), 13301–13306. doi:10.1073/ pnas.1411529111

Sankararaman, S., Mallick, S., Dannemann, M., Prüfer, K., Kelso, J., Pääbo, S., et al. (2014). The genomic landscape of Neanderthal ancestry in present-day humans. *Nature* 507, (7492), 354. doi:10.1038/nature12961

Sykes, B. (2001). *The Seven Daughters of Eve: the science that reveals our genetic ancestry.* New York: Norton.

Hublin, Jean-Jacques. Interview, November 2014.

Prüfer, Kay. Interview, September 2013.

Pääbo, Svante. Interview, November 2014.

Stefánsson, Kári. Interview, 1998.

The Flute Players

I strongly recommend going on an Ice Age safari to the Swabian uplands to view the world's oldest known musical instruments and the world's oldest figurative art. However, it is worth bearing in mind that Germany is a highly decentralised federal republic without any central authority to make life easier for Ice Age tourists.

The most important finds are dispersed; some can be found at the museum in Ulm, others at the castle museum in Tübingen and in the little town of Blaubeuren. The small museum in Blaubeuren is particularly pleasant, with a special room where you can listen to the flutes, and an exhibition that shows what the flute players may have looked like, and how their gear differed from that of the Neanderthals.

It is only a few kilometres from Blaubeuren to the caves at Geißenklösterle and Hohle Fels. From Ulm, you can take the train to Schelklingen, after which it is no more than a half-hour walk. The bus from Blaubeuren to Ehingen takes you even closer; get off at the bus stop marked 'Hohle Fels'.

Geißenklösterle can only be viewed from the outside, and you have to scale a steep slope to reach it. Hohle Fels, however, is more accessible, and it is sometimes open to the public. For further information, it is best to send an email in advance to Erwin Haggenmüller (e.haggenmueller@t-online.de), Dieter Frey (dodifrey@web.de) or info@schelklingen.de.

Bäckman, L. & Hultkrantz, Å. (1978). *Studies in Lapp Shamanism.* Stockholm: Almqvist & Wiksell International.

Conard, N. (2003). Palaeolithic ivory sculptures from southwestern Germany and the origins of figurative art. *Nature, 426* (6968), 830–832.

Conard, N. (2009). A female figurine from the basal Aurignacian of Hohle Fels Cave in southwestern Germany. *Nature, 459* (7244), 248–252. doi:10.1038/nature07995

Conard, N. J. & Bolus, M. (2003). Radiocarbon dating the appearance of modern humans and timing of cultural innovations in Europe: new results and new challenges. *Journal of Human Evolution, 44,* 331–371. doi:10.1016/S0047-2484(02)00202-6

Conard, N. J., Malina, M., & Münzel, S. C. (2009). New flutes document the earliest musical tradition in southwestern Germany. *Nature, 460* (7256), 737–740. doi:10.1038/nature08169

Cook, J. (2013). *Ice Age Art: arrival of the modern mind.* London: British Museum Press.

Eichmann, R., Jianjun, F., & Koch, L-F. (2015). *Studien zur Musikarchäologie X.* In press.

Fagan, B. M. & Durrani, N. (2013). *People of the Earth: an introduction to world prehistory.* Boston: Pearson.

Fu, Q., Li, H., Moorjani, P., Jay, F., Slepchenko, S. M., Bondarev, A. A., et al. (2014). Genome sequence of a 45,000-year-old modern human from western Siberia. *Nature, 514* (7523), 445–449. doi:10.1038/nature13810

Higham, T., Basell, L., Jacobi, R., Wood, R., Ramsey, C., & Conard, N. (2012). Testing models for the beginnings of the Aurignacian and the advent of figurative art and music: the radiocarbon chronology of Geißenklösterle. *Journal of Human Evolution, 62* (6), 664–676. doi:10.1016/j.jhevol.2012.03.003

Karlsson, J. (1970). A double dominant genetic mechanism for schizophrenia. *Hereditas, 65* (2), 261–267.

Kong, A., Frigge, M. L., Masson, G., Besenbacher, S., Sulem, P., Magnusson, G., et al. (2012). Rate of *de novo* mutations and the importance of father's age to disease risk. *Nature, 488* (7412), 471–475. doi:10.1038/nature11396

Kyaga, S., Lichtenstein, P., Boman, M., Hultman, C., Långström, N., & Landén, M. (2011). Creativity and mental disorder: family study of 300,000 people with severe mental disorder. *The British Journal of Psychiatry, 199* (5), 373–379. doi:10.1192/bjp. bp.110.085316

Lewis-Williams, J. D. (2004[2002]). *The Mind in the Cave: consciousness and the origins of art.* London: Thames & Hudson.

Morwood, M. J., Aubert, M. M., Brumm, A. A., Ramli, M. M., Sutikna, T. T., Saptomo, E. W., et al. (2014). Pleistocene cave art from Sulawesi, Indonesia. *Nature, 514* (7521), 223–227. doi:10.1038/nature13422

Conard, Nicholas. Interview, September 2013.

Cook, Jill. Interview, March 2013.

Lichtenstein, Paul. Interview, 2014.

Taller, Andreas. Conversation, September 2013.

Blaubeuren Museum of Prehistory (Urgeschichtliches Museum Blaubeuren). Visited in September 2013.

Geißenklösterle. Visit, September 2013.

Hohentübingen Castle Museum (Museum Schloss Hohentübingen). Visited in 2013.

Hohle Fels. Visit, September 2013.

First on the Scene in Europe

I received a great deal of help on the Kostenki site from the Russian archaeologist Andrei Sinitsyn. John F. Hoffecker, an American expert on eastern European archaeology, was also very helpful. Unfortunately, I have not yet been able to visit either the local museum in Kostenki or the Kunstkamera (the Peter the Great Museum of Anthropology and Ethnography) in St Petersburg, where the photo of the model was taken.

Anikovich, M., Sinitsyn, A., Hoffecker, J., Holliday, V., Popov, V., Lisitsyn, S., et al. (2007). Early Upper Paleolithic in Eastern Europe and implications for the dispersal of modern humans. *Science*, *315* (5809), 223–226. doi:10.1126/science.1133376

Benazzi, S., Douka, K., Fornai, C., Bauer, C., Kullmer, O., Svoboda, J., et al. (2011). Early dispersal of modern humans in Europe and implications for Neanderthal behaviour. *Nature*, *479* (7374), 525–528. doi:10.1038/nature10617

Benazzi, S., Bailey, S. E., Peresani, M., Mannino, M. A., Romandini, M., Richards, M. P., & Hublin, J. (2014). Middle Paleolithic and Uluzzian human remains from Fumane Cave, Italy. *Journal of Human Evolution*, *70*, 61–68. doi:10.1016/j.jhevol.2014.03.001

Dalén, L., Orlando, L., Shapiro, B., Brandström-Durling, M., Quam, R., Gilbert, M., et al. (2012). Partial genetic turnover in Neanderthals: continuity in the East and population replacement in the West. *Molecular Biology and Evolution*, *29* (8), 1893–1897. doi:10.1093/molbev/mss074

Douka, K., Higham, T. G., Wood, R., Boscato, P., Gambassini, P., Karkanas, P., et al. (2014). On the chronology of the Uluzzian. *Journal of Human Evolution*, *68*, 1–13. doi:10.1016/j.jhevol.2013.12.007

Higham, T., Compton, T., Stringer, C., Jacobi, R., Shapiro, B., Trinkaus, E., et al. (2011). The earliest evidence for anatomically modern humans in northwestern Europe. *Nature, 479* (7374), 521–524. doi:10.1038/nature10484

Higham, T., Douka, K., Wood, R., Ramsey, C., Brock, F., Basell, L., et al. (2014). The timing and spatiotemporal patterning of Neanderthal disappearance. *Nature, 512* (7514), 306–309. doi:10.1038/nature13621

Hublin, J. (2012). The earliest modern human colonization of Europe. *Proceedings of the National Academy of Sciences of the United States of America, 109* (34), 13471–13472. doi:10.1073/pnas.1211082109

Krause, J., Briggs, A., Kircher, M., Maricic, T., Zwyns, N., Derevianko, A., & Pääbo, S. (2010). A complete mtDNA genome of an early modern human from Kostenki, Russia. *Current Biology, 20* (3), 231–236. doi:10.1016/j.cub.2009.11.068

Maricic, T., Günther, V., Georgiev, O., Gehre, S., Ćurlin, M., Schreiweis, C., et al. (2013). A recent evolutionary change affects a regulatory element in the human *FOXP2* gene. *Molecular Biology and Evolution, 30* (4), 844.

Marom, A., McCullagh, J., Higham, T., Sinitsyn, A., & Hedges, R. (2012). Single amino acid radiocarbon dating of Upper Paleolithic modern humans. *Proceedings of the National Academy of Sciences of the United States of America, 109* (18), 6878–6881. doi:10.1073/pnas.1116328109

Oppenheimer, C. (2011). *Eruptions that Shook the World [Electronic resource].* Cambridge: Cambridge University Press.

Trinkaus, E., Zilhão, J., & Constantin, S. (2013). *Life and Death at the Peştera cu Oase: a setting for modern human emergence in Europe.* Oxford: Oxford University Press.

Wood, R., Barroso-Ruíz, C., Caparrós, M., Jordá Pardo, J., Galván Santos, B., & Higham, T. (2013). Radiocarbon dating casts doubt on the late chronology of the Middle to Upper Palaeolithic transition in southern Iberia. *Proceedings of the National Academy of Sciences of the United States of America, 110* (8), 2781–2786. doi:10.1073/pnas.1207656110

Conard, Nicholas. Interview, 2013.

Higham, Thomas. Email, 2014.

Hoffecker, John. Email, 2014.

Hublin, Jean-Jacques. Interview, November 2014. Email, December 2014.

Pääbo, Svante. Email, May 2015.

Sinitsyn, Andrei. Email, 2014.

Trinkhaus, Erik. Email, 2014.
Wohlfarth, Barbara. Interview, July 2013.
Musée National de Préhistoire, Les Eyzies-de-Tayac. Visited in 2013.

Mammoths in Brno

Brno, just over an hour by train from Vienna, is definitely worth a visit. Anthropos is among Europe's more informative prehistory museums, and a visit to the monastery that was once the workplace of Gregor Mendel, the father of genetics, is surely part of our general education now that DNA technology is developing at such a dramatic rate.

And don't miss Brno's main square, where a large farmers' market is held. This is particularly inviting in September, with cups of newly made wine on sale at market stalls.

Cook, J. (2013). *Ice Age Art: arrival of the modern mind.* London: British Museum Press.
Fu, Q., Mittnik, A., Johnson, P., Bos, K., Lari, M., Bollongino, R., et al. (2013). A revised timescale for human evolution based on ancient mitochondrial genomes. *Current Biology, 23* (7), 553–559. doi:10.1016/j.cub.2013.02.044
Cook, Jill. Interview, March 2013.
Svoboda, Jiři. Interview, September 2013. Email, 2014.
Anthropos Museum, Brno. Visited in September 2013.

Cro-Magnon

If you choose to visit just one of the destinations described in this book, make it Les-Eyzies-de-Tayac. The French state and the local authorities have done everything they can here to pool their resources and make sure Ice Age tourists get the most out of their visit. Many of the sites are just 10 minutes' walk apart.

It is best to avoid the high season in July and August, when the hotels are full and the narrow streets heaving with cars.

The train journey from Bordeaux takes about three hours. It is about five minutes' walk from the little station to Cro-Magnon, and just a few minutes more to Abri Pataud, which has a small museum of its own. The big National Museum of Prehistory, the *Musée National de Préhistoire*, is a few hundred metres further on. Tremendously ambitious, the museum places considerable demands on the visitor. It is advisable to go on a guided tour, which will put everything into context.

The museum's contact details are available on its homepage. It is open throughout the week in the high season; during the rest of the year it closes on Tuesdays.

The expert guides at the *Musée National de Préhistoire*, such as Florence Landais, are responsible for coordinating most things to do with prehistory in the region. However, this does not apply to Abri Pataud, which is covered by another state organisation, or to Rouffignac, which is privately owned.

The *Pôle International de la Préhistoire* (International Prehistory Centre) occupies a large new building near the museum. Its facilities include an extensive collection of books, magazines and films, which visitors can use free of charge. The centre is closed on Saturdays.

Font-de-Gaume is about half an hour on foot from Les Eyzies. Tickets can be purchased on the spot on the same day as the tour. It's best to be there by 9.30 a.m., when they open! In the same direction, but a little further away, are Les Combarelles and Abri du Cap Blanc (the latter can be reached by bike or car, rather than on foot).

Abri du Poisson, Laugerie Haute and some other sites are about half an hour's walk from Les Eyzies in the other direction.

The simplest way to get to Rouffignac is to hire a car.

The tourist office (*office de tourisme*) in Les Eyzies can put visitors in touch with car rental firms and hire out cycles.

The Department of Geological Sciences at the University of Stockholm was very helpful when I was planning my trip to Les Eyzies. Barbara Wohlfarth and Otto Hermelin used to organise trips there as part of the introductory course on human evolution. Sadly, the course is no longer running.

It was on Barbara Wohlfarth's advice that I decided against visiting the most famous cave in the region, Lascaux. The public can no longer enter the cave, but have to make do with copies of the paintings on plastic backgrounds. Even so, it seems there are too many visitors for comfort.

The cave at Pech Merle, about two hours by car from Les Eyzies, is said to be extraordinary. Unfortunately, I was unable to fit in a visit.

Staying at the Cro-Magnon guesthouse was interesting and enjoyable, but rather expensive. More affordable alternatives are chalets at the local campsite or the Auberge de Musée. I ate best at Hôtel Les Glycines, and the Café-Brasserie de la Mairie at the foot of the big prehistory museum also served tasty dishes at affordable prices.

Achilli, A., Rengo, C., Battaglia, V., Pala, M., Olivieri, A., Fornarino, S., et al. (2005). Saami and Berbers – an unexpected mitochondrial DNA link. *The American Journal of Human Genetics*, 76 (5), 883–886.

Bäckman, L. & Hultkrantz, Å. (1978). *Studies in Lapp Shamanism*. Stockholm: Almqvist & Wiksell International.

Clottes, J. (2011[2008]). *Cave Art*. New York: Phaidon Press.

Desdemaines-Hugon, C. & Tattersall, I. (2010). *Stepping-stones: a journey through the Ice Age caves of the Dordogne*. New Haven: Yale University Press.

Fu, Q., Mittnik, A., Johnson, P., Bos, K., Lari, M., Bollongino, R., et al. (2013). A revised timescale for human evolution based on ancient mitochondrial genomes. *Current Biology*, 23 (7), 553–559. doi:10.1016/j.cub.2013.02.044

Lewis-Williams, J. D. (2004[2002]). *The Mind in the Cave: consciousness and the origins of art*. London: Thames & Hudson.

Tambets, K., Rootsi, S., Kivisild, T., Help, H., Serk, P., Loogväli, E., et al. (2004). The western and eastern roots of the Saami – the story of genetic "outliers" told by mitochondrial DNA and Y chromosomes. *The American Journal of Human Genetics*, 74 (4), 661–682. doi:10.1086/383203

Wohlfarth, Barbara, University of Stockholm. Excursion guide for the 2013 course on human evolution.

Berlin, Rolf. Email, August 2011.

Chiotti, Laurent. Interview, September 2013.

Krause, Johannes. Interview, September 2013.

Plassard, Frédéric. Interview, September 2013.

Svoboda, Jiři. Interview, September 2013.

Abri du Cap Blanc, Les Eyzies-de-Tayac. Visited in September 2013.

Abri Pataud and Laugerie Haute, Les Eyzies-de-Tayac. Visited in September 2013.

Font-de-Gaume, Les Eyzies-de-Tayac. Visited in September 2013.

Les Combarelles, Les Eyzies-de-Tayac. Visited in September 2013.

Musée de l'Abri Pataud, Les Eyzies-de-Tayac. Visited in September 2013.

Musée National de Préhistoire, Les Eyzies-de-Tayac. Visited in September 2013.

Rouffignac. Visited in September 2013.

The First Dog

See the chapter on the Falbygden area for information on the museum.

Freedman, A., Gronau, I., Schweizer, R., Ortega-Del Vecchyo, D., Han, E., Silva, P., et al. (2014). Genome sequencing highlights the dynamic early history of dogs. *PLOS Genetics*, *10* (1), e1004016. doi:10.1371/journal.pgen.1004016

Fu, Q., Mittnik, A., Johnson, P., Bos, K., Lari, M., Bollongino, R., et al. (2013). A revised timescale for human evolution based on ancient mitochondrial genomes. *Current Biology*, *23* (7), 553–559. doi:10.1016/j.cub.2013.02.044

Giemsch, L. (2014). *Eiszeitjäger: Leben im Paradies: Europa vor 15000 Jahren*. Mainz: Nünnerich-Asmus.

Pettitt, P. (2011). *The Palaeolithic Origins of Human Burial*. Abingdon: Routledge.

Skoglund, P., Ersmark, E., Palkopoulou, E., & Dalén, L. (2015). Ancient wolf genome reveals an early divergence of domestic dog ancestors and admixture into high-latitude breeds. *Current Biology*, doi:10.1016/j.cub.2015.04.019

Thalmann, O., Shapiro, B., Cui, P., Schuenemann, V., Sawyer, S., Greenfield, D., et al. (2013). Complete mitochondrial genomes of ancient canids suggest a European origin of domestic dogs. *Science*, *342* (6160), 871–874. doi:10.1126/science.1243650

Schmitz, Ralph W. Email, 2014.

Street, Martin. Email, 2014.

Thalmann, Olaf. Email, 2014.

Doggerland

Research into Doggerland is very recent, and there is surprisingly little information available. For those who want to know more, I can only recommend Vincent Gaffney's book and a few display cases at the Danish National Museum in Copenhagen.

This museum is the best resource in Scandinavia for anyone wanting to learn about the Palaeolithic in northern Europe (as well as the Neolithic period, the Bronze Age and the Iron Age). Moreover, entrance is free, and the area around the museum has some pleasant restaurants serving smørrebrød, the typical Danish open sandwiches.

Coles, B. (2000). *Doggerland: The cultural dynamics of a shifting coastline.* London: Geological Society. doi:10.1144/GSL.SP.2000.175.01.27

Fu, Q., Rudan, P., Pääbo, S., & Krause, J. (2012). Complete mitochondrial genomes reveal Neolithic expansion into Europe. *PLOS ONE,* 7 (3), 1–6. doi:10.1371/journal.pone.0032473

Gaffney, V. L., Fitch, S., & Smith, D. N. (2009). *Europe's Lost World: the rediscovery of Doggerland.* York: Council for British Archaeology.

Gaffney, Vincent. Interview, January 2015.

Wohlfarth, Barbara. Interview, March 2014.

The Ice Age Ends

The idea that people witnessed the sudden appearance of a huge waterfall to the north of Mount Billingen and the time when the water in the Öresund Strait ran dry is, of course, an invention. We cannot know whether there were any eyewitnesses. I borrowed most of the fictional description from Professor Svante Björck of Lund University, Sweden's main expert on these geological events.

The finds from Huseby Klev are on show in a small local museum at Orust, which is open for a few weeks in summer and by prior arrangement.

The island of Stora Karlsö is a wonderful place to visit, especially at the beginning of summer, when the orchids are in bloom and the common murre (guillemot) chicks are leaping from the cliffs. There are daily boat trips until the end of August. The Stora Förvar cave, near the harbour, is freely accessible to the public.

The Barum woman is exhibited at the Swedish History Museum in Stockholm. In addition to her skeleton, there is a whole room full of information about the environment in which she lived.

Apart from these sites, the best and most complete information about the Scandinavian Stone Age is to be found at the Danish National Museum in Copenhagen.

Andersson, M. & Knarrström, B. (1999). *Senpaleolitikum i Skåne: en studie av materiell kultur och ekonomi hos Sveriges första fångstfolk.* Stockholm: Swedish National Heritage Board, Archaeological Research Department.

Bailey, G. N. & Spikins, P. (eds) (2008). *Mesolithic Europe.* Cambridge: Cambridge University Press.

Bjerck, H. B. (2009). Colonizing seascapes: comparative perspectives on the development of maritime relations in Scandinavia and Patagonia. *Arctic Anthropology*, *46* (1–2), 118–131.

Carlson, A. (2013). The Younger Dryas climate event. *Encyclopedia of Quaternary Science*, *3*, 126–134. doi:10.1016/B978-0-444-53643-3.00029-7

Hernek, R. & Nordqvist, B. (1995). *Världens äldsta tuggummi?Ett urval spännande arkeologiska fynd och upptäckter som gjordes vid Huseby klev, och andra platser, inför väg 178 över Orust* (1st edition). Kungsbacka: Byrån för arkeologiska undersökningar, Swedish National Heritage Board.

IPCC. (2013). *Climate Change 2013: the physical science basis*. Chapter 5: Information from Paleoclimate Archives. Cambridge and New York: Cambridge University Press.

Lindqvist, C. & Possnert, G. (1999). The first seal hunter families on Gotland: on the Mesolithic occupation in the Stora Förvar cave. *Current Swedish Archaeology*, *7*, 65–87.

Malyarchuk, B. (2008). Mitochondrial DNA phylogeny in Eastern and Western Slavs. *Molecular Biology and Evolution*, *25* (8), 1651–1658.

Malyarchuk, B., Derenko, M., Grzybowski, T., Perkova, M., Rogalla, U., Vanecek, T., & Tsybovsky, I. (2010). The peopling of Europe from the mitochondrial haplogroup U5 perspective. *PLOS ONE*, *5* (4), 1–6. doi:10.1371/journal.pone.0010285

Masson-Delmotte, V., Swingedouw, D., Landais, A., Seidenkrantz, M., Gauthier, E., Bichet, V., et al. (2012). Greenland climate change: from the past to the future. *Wiley Interdisciplinary Reviews: Climate Change*, *3* (5), 427–449.

Nordqvist, B. (2005). *Huseby klev: en kustboplats med bevarat organiskt material från äldsta mesoliticum till järnålder: Bohuslän, Morlanda socken, Huseby 2:4 och 3:13, RAÄ 89 och 485: arkeologisk förundersökning och undersökning*. Mölndal: Contract Archaeological Service (West), Swedish National Heritage Board, Archaeological Research Department.

Andersson, Magnus. Interview, February 2014.

Andrén, Thomas. Telephone interview, February 2014.

Apel, Jan. Telephone interview, November 2014 and March 2015.

Bjerck, Hein Bjartmann. Email, March 2014.

Björck, Svante. Interview, March 2014.

Broström, Anna. Interview, February 2014.

Jakobsson, Mattias. Interviews, November and December 2014.

Jennbert, Kristina. Interview, February 2014.

Magnell, Ola. Interview, February 2014.

Nordqvist, Bengt. Telephone interview, March 2014.

Price, Douglas. Interview, February 2014.
Storå, Jan. Telephone interview, December 2014.
Wohlfarth, Barbara. Interview, March 2014.

Dark Skin, Blue Eyes

Cerqueira, C., Paixão-Côrtes, V., Zambra, F., Salzano, F., Hünemeier, T., & Bortolini, M. (2012). Predicting homo pigmentation phenotype through genomic data: from Neanderthal to James Watson. *American Journal of Human Biology*, *24* (5), 705–709. doi:10.1002/ajhb.22263

Draus-Barini, J., Walsh, S., Pośpiech, E., Kupiec, T., Głab, H., Branicki, W., & Kayser, M. (2013). Bona fide colour: DNA prediction of human eye and hair colour from ancient and contemporary skeletal remains. *Investigative Genetics*, *4* (1), doi:10.1186/2041-2223-4-3

Fortes, G., Speller, C., Hofreiter, M., & King, T. (2013). Phenotypes from ancient DNA: approaches, insights and prospects. *Bioessays*, *35* (8), 690–695.

Gaffney, V. L., Fitch, S., & Smith, D. N. (2009). *Europe's Lost World: the rediscovery of Doggerland*. York: Council for British Archaeology.

Gamba, C., Jones, E. R., Teasdale, M. D., McLaughlin, R. L., Gonzalez-Fortes, G., Mattiangeli, V., et al. (2014). Genome flux and stasis in a five millennium transect of European prehistory. *Nature Communications*, *5* (10), 5257. doi:10.1038/ncomms6257

Haak, W., Lazaridis, I., Patterson, N., Rohland, N., Mallick, S., Llamas, B., et al. (2015). Massive migration from the steppe was a source for Indo-European languages in Europe. *Nature*, *522* (7555), 207–211. doi:10.1038/nature14317

Ingram, C. E., Mulcare, C. A., Itan, Y., Thomas, M. G., & Swallow, D. M. (2009). Lactose digestion and the evolutionary genetics of lactase persistence. *Human Genetics*, *124* (6), 579. doi:10.1007/s00439-008-0593-6

Itan, Y., Jones, B., Ingram, C., Swallow, D., & Thomas, M. (2010). A worldwide correlation of lactase persistence phenotype and genotypes. *BMC Evolutionary Biology*, *10* (1), doi:10.1186/1471-2148-10-36

Lazaridis, I., Patterson, N., Mittnik, A., Renaud, G., Mallick, S., Kirsanow, K., et al. (2014). Ancient human genomes suggest three ancestral populations for present-day Europeans. *Nature*, *513* (7518), 409–413. doi:10.1038/nature13673

Mathiesen, I. (2015). Eight thousand years of natural selection in Europe. *bioRxiv* preprint, first published online on 14 March 2015.

Olalde, I., Allentoft, M., Sanchez-Quinto, F., Santpere, G., Chiang, C., DeGiorgio, M., et al. (2014). Derived immune and ancestral pigmentation alleles in a 7,000-year-old Mesolithic European. *Nature*, 507 (7491), 225–228.

Sten, S. (ed.) (2000). Barumkvinnan: nya forskningsrön. *Fornvännen*, 95 (2000), 73–87.

Ahlström, Torbjörn. Telephone interview, March 2014.

Hallgren, Fredrik. Interview, March 2014.

Sten, Sabine. Telephone interview, March 2014.

Swedish History Museum, Stockholm. Visited several times between 2013 and 2015.

Climate and Forests

Gaffney, V. L., Fitch, S., & Smith, D. N. (2009). *Europe's Lost World: the rediscovery of Doggerland*. York: Council for British Archaeology.

Lazaridis, I., Patterson, N., Mittnik, A., Renaud, G., Mallick, S., Kirsanow, K., et al. (2014). Ancient human genomes suggest three ancestral populations for present-day Europeans. *Nature*, 513 (7518), 409–413. doi:10.1038/nature13673

Olalde, I., Allentoft, M., Sánchez-Quinto, F., Santpere, G., Chiang, C., Degiorgio, M., et al. (2014). Derived immune and ancestral pigmentation alleles in a 7,000-year-old Mesolithic European. *Nature*, 507, 225–228. doi:10.1038/nature12960

Weninger, B. (2008). The catastrophic final flooding of Doggerland by the Storegga Slide tsunami. *Documenta Praehistorica*, 35, 1–24.

Björck, Svante. Telephone interview, March 2014.

Gaffney, Vincent. Interview, January 2015.

Wohlfarth, Barbara. Interview, March 2014.

Am I a Sami?

At the very last moment, just before it was time to hand in the manuscript, I decided to take the night train up to the Sami Museum, Ájtte, in Jokkmokk. I am very glad I did. Ájtte is an outstandingly good museum and an important part of Sweden's history.

Broberg, G. & Tydén, M. (1991). *Oönskade i folkhemmet: rashygien och sterilisering i Sverige*. Stockholm: Gidlund.

Ingman, M. & Gyllensten, U. (2007). A recent genetic link between Sami and the Volga–Ural region of Russia. *European Journal of Human Genetics*, *15* (1), 115–120. doi:10.1038/sj.ejhg.5201712

Janson, T. (2013). *Germanerna: myten, historien, språken*. Stockholm: Norstedts.

Krebs, C. B. (2011). *A Most Dangerous Book: Tacitus's Germania from the Roman Empire to the Third Reich* (1st edition). New York: W. W. Norton & Co.

Malyarchuk, B., Derenko, M., Grzybowski, T., Perkova, M., Rogalla, U., Vanecek, T., & Tsybovsky, I. (2010). The peopling of Europe from the mitochondrial haplogroup U5 perspective. *PLOS ONE*, *5* (4), 1–6. doi:10.1371/journal.pone.0010285

Manker, E. (1947). *De svenska fjällapparna*. Stockholm: Svenska turistföreningens förlag.

Tambets, K., Rootsi, S., Kivisild, T., Help, H., Serk, P., Loogväli, E., et al. (2004). The western and eastern roots of the Saami – the story of genetic "outliers" told by mitochondrial DNA and Y chromosomes. *The American Journal of Human Genetics*, *74*, 661–682. doi:10.1086/383203

Aronsson, Kjell-Åke. Interview, December 2014.

Gyllensten, Ulf. Interview, December 2014.

Tambets, Kristiina. Email, December 2014.

Tonnesen, Gail. Email, 2013 and 2014.

Pottery Makes its Appearance

Craig, O., Saul, H., Lucquin, A., Nishida, Y., Taché, K., Clarke, L., et al. (2013). Earliest evidence for the use of pottery. *Nature*, *496* (7445), 351–354. doi:10.1038/nature12109

Marzurkevich, A. & Dolbunova, E. (2012). The oldest pottery and Neolithisation of Eastern Europe. *Fontes Archaeologici Poznaniensis*, *48*.

Saul, H., Madella, M., Fischer, A., Glykou, A., Hartz, S., & Craig, O. (2013). Phytoliths in pottery reveal the use of spice in European prehistoric cuisine. *PLOS ONE*, *8* (8), e70583. doi:10.1371/journal.pone.0070583

Swedish National Heritage Board. Contract Archaeology Service (South) (2004). *Stone Age Scania: significant places dug and read by contract archaeology* (1st edition). Stockholm: Riksanti-kvarieämbetets förlag.

Wu, X., Zhang, C., Goldberg, P., Cohen, D., Pan, Y., Arpin, T., & Bar-Yosef, O. (2012). Early pottery at 20,000 years ago in

Xianrendong Cave, China. *Science*, *336* (6089), 1696–1700. doi:10.1126/science.1218643

Dolbunova, Ekaterina. Lecture, September 2013. Email, September 2013.

Hallgren, F. Interview, March 2014.

The Farmers Arrive

Burenhult, G. (ed.) (2010). *Arkeologi i Norden. 1.* Stockholm: Natur & Kultur.

Jennbert, K. (1984). *Den produktiva gåvan: tradition och innovation i Sydskandinavien för omkring 5300 år sedan.* Diss. Lund: Univ. Bonn.

Lazaridis, I., Patterson, N., Mittnik, A., Renaud, G., Mallick, S., Kirsanow, K., et al. (2014). Ancient human genomes suggest three ancestral populations for present-day Europeans. *Nature, 513* (7518), 409–413. doi:10.1038/nature13673

Nordic Museum, Lagersberg Foundation (1998). *Det svenska jordbrukets historia. [Bd 1], Jordbrukets första femtusen år: [4000 f. Kr.–1000 e. Kr.].* Stockholm: Natur & Kultur/LT in conjunction with the Nordic Museum, Stockholm, and the Lagersberg Foundation.

Skoglund, P., Malmström, H., Raghavan, M., Storå, J., Hall, P., Willerslev, E., Gilbert, M. T. P., Götherström, A., & Jakobsson, M. (2012). Origins and genetic legacy of Neolithic farmers and hunter-gatherers in Europe. *Science, 336,* 466–469. doi:10.1126/science.1216304

Sørensen, L. & Karg, S. (2014). The expansion of agrarian societies towards the north – new evidence for agriculture during the Mesolithic/Neolithic transition in Southern Scandinavia. *Journal of Archaeological Science, 51,* 98–114. doi:10.1016/j.jas.2012.08.042

Gronenborn, Detlef. Interview, September 2013.

Haak, Wolfgang. Email, July 2013.

Jennbert, Kristina. Interview, February 2014.

Price, Douglas. Interview, February 2014.

The Swedish Research Council. Email, December 2013.

Family Tree DNA. Personal analysis, 2013.

Jakobsson, M., Götherström, A., Storå, J., Skoglund, P., Burenhult, G. Press conference in Uppsala, April 2012.

The Swedish Foundation for Humanities and Social Sciences. Press release, October 2013.

University of Stockholm. Press release, October 2013.

University of Uppsala. Press release, October 2013.

Syria

Fernández, E., Pérez-Pérez, A., Gamba, C., Prats, E., Cuesta, P., Anfruns, J., et al. (2014). Ancient DNA analysis of 8000 B.C. Near Eastern farmers supports an Early Neolithic pioneer maritime colonization of mainland Europe through Cyprus and the Aegean Islands. *PLOS Genetics*, *10* (6), e1004401. doi:10.1371/journal.pgen.1004401

The Boat to Cyprus

Exploring the origins of agriculture in depth could scarcely be made any simpler or more comfortable. Tourism being Cyprus's main economic activity, there are hotels, restaurants and museums galore. You can get around fairly easily using ordinary buses, and where they are lacking there are taxis, cycles for hire and rental cars. The most informative destination as regards early agriculture is the World Heritage Site of Khirokitia, near the A1 national highway a few dozen kilometres west of Larnaca. A regular bus service will take you nearly all the way to the site.

The Cyprus Museum in the capital, Nicosia, is also quite informative about the first farmers. Look at the display cabinets in the first room.

Driscoll, C., Menotti-Raymond, M., Roca, A., Hupe, K., Johnson, W., Geffen, E., & Macdonald, D. (2007). The Near Eastern origin of cat domestication. *Science*, *317* (5837), 519–523.

Kuijt, I. & Finlayson, B. (2009). Evidence for food storage and predomestication granaries 11,000 years ago in the Jordan Valley. *Proceedings of the National Academy of Sciences of the United States of America*, *106* (27), 10966–10970. doi:10.1073/pnas.0812764106

Málek, J. (2006). *The Cat in Ancient Egypt*. London: British Museum.

Montague, M. J., Gang, L., Gandolfi, B., Khan, R., Aken, B. L., Searle, S. J., et al. (2014). Comparative analysis of the domestic cat genome reveals genetic signatures underlying feline biology and domestication. *Proceedings of the National Academy of Sciences of the United States of America*, *111* (48), 17230–17235. doi:10.1073/pnas.1410083111

Van Neer, W., Linseele, V., Friedman, R., & De Cupere, B. (2014). More evidence for cat taming at the Predynastic elite cemetery of Hierakonpolis (Upper Egypt). *Journal of Archaeological Science*, *45*, 103–111. doi:10.1016/j.jas.2014.02.014

Vigne, J., Guilaine, J., Debue, K., Haye, L., & Gérard, P. (2004). Early taming of the cat in Cyprus. *Science, 304* (5668), 259.

Vigne, J., Briois, F., Zazzo, A., Willcox, G., Cucchi, T., Thiébault, S., et al. (2012). First wave of cultivators spread to Cyprus at least 10,600 years ago. *Proceedings of the National Academy of Sciences of the United States of America, 109* (22), 8445–8449. doi:10.1073/pnas.1201693109

Häggman, Sofia. Interview, May 2014.

Johansson, Carolin. Email, December 2014.

McCartney, Carole. Interview and field visit, April 2014.

Mühlenbock, Christian. Conversation, May 2015.

Vigne, Jean-Denis. Email, April 2014.

Visit to Cyprus, April 2014.

The First Beer

There are organised tours of Göbekli Tepe for tourists.

Bar-Yosef, O. (2011). Climatic fluctuations and early farming in West and East Asia. *Current Anthropology, 52* (S4), doi:10.1086/659784

Colledge, S. & Conolly, J. (2007). *The Origins and Spread of Domestic Plants in Southwest Asia and Europe [Electronic resource].* California: Left Coast Press, Inc.

Dietrich, O., Heun, M., Notroff, J., Schmidt, K., Zarnkow, M., & Carver, M. (2012). The role of cult and feasting in the emergence of Neolithic communities. New evidence from Göbekli Tepe, south-eastern Turkey. *Antiquity, 86* (333), 674–695.

Heun, M., Schäfer-Pregl, R., Klawan, D., Castagna, R., Accerbi, M., Borghi, B., & Salamini, F. (1997). Site of einkorn wheat domestication identified by DNA fingerprinting. *Science, 278* (5341), 1312–1314.

Larson, G., Piperno, D. R., Allaby, R. G., Purugganan, M. D., Andersson, L., Arroyo-Kalin, M., et al. (2014). Current perspectives and the future of domestication studies. *Proceedings of the National Academy of Sciences of the United States of America, 111* (17), 6139–6146. doi:10.1073/pnas.1323964111

Luo, M., Yang, Z., You, F., Kawahara, T., Waines, J., & Dvorak, J. (2007). The structure of wild and domesticated emmer wheat populations, gene flow between them, and the site of emmer domestication. *Theoretical and Applied Genetics, 114* (6), 947–959.

Piperno, D., Weiss, E., Holst, I., & Nadel, D. (2004). Processing of wild cereal grains in the Upper Palaeolithic revealed by starch grain analysis. *Nature, 430* (7000), 670–673.

Price, T. & Bar-Yosef, O. (2011). The origins of agriculture: new data, new ideas. An Introduction to Supplement 4. *Current Anthropology, 52* (S4), doi:10.1086/659964

Vigne, J., Carrère, I., Briois, F., & Guilaine, J. (2011). The early process of mammal domestication in the Near East: new evidence from the Pre-Neolithic and Pre-Pottery Neolithic in Cyprus. *Current Anthropology, 52* (S4), doi:10.1086/659306

Vigne, J., Briois, F., Zazzo, A., Willcox, G., Cucchi, T., Thiébault, S., et al. (2012). First wave of cultivators spread to Cyprus at least 10,600 y ago. *Proceedings of the National Academy of Sciences of the United States of America, 109* (22), 8445–8449. doi:10.1073/pnas.1201693109

Weiss, E. & Zohary, D. (2011). The Neolithic Southwest Asian founder crops, their biology and archaeobotany. *Current Anthropology, 52* (S4), doi:10.1086/658367

Willcox, G. & Herveux, L. (2013). Late Pleistocene/Early Holocene charred plant remains: preliminary report. In R. F. Mazurowski, Y. Kanjou (eds), *Tell Qaramel 1999–2007. Protoneolithic and Early Pre-pottery Neolithic Settlement in Northern Syria.* 120/130. Warsaw: PCMA.

Zeder, M. A. (2011). The origins of agriculture in the Near East. *Current Anthropology, 52* (S4), doi:10.1086/659307

Dietrich, Oliver. Email, October 2014.

McCartney, Carole. Interview and field visit, March 2014.

Vigne, Jean-Denis. Email, March 2014.

The Farmers' Westward Voyages

Badro, D. A., Douaihy, B., Haber, M., Youhanna, S. C., Salloum, A., Ghassibe-Sabbagh, M., et al. (2013). Y-chromosome and mtDNA genetics reveal significant contrasts in affinities of modern Middle Eastern populations with European and African populations. *PLOS ONE, 8* (1), 1–11. doi:10.1371/journal.pone.0054616

Bengtson, J. D. (2009). Basque and the other Mediterranean languages. *Mother Tongue, 14,* 157–175.

Huysecoml, E., Rasse, M., Lespez, L., Neumann, K., Fahmy, A., Ballouche, A., et al. (2009). The emergence of pottery in Africa during the tenth millennium cal BC: new evidence from Ounjougou (Mali). *Antiquity, 83* (322), 905–917.

Manco, J. (2013). *Ancestral Journeys: the peopling of Europe from the first ventures to the Vikings.* London: Thames & Hudson.

Rowley-Conwy, P. (2011). Westward Ho! The spread of agriculture from Central Europe to the Atlantic. *Current Anthropology, 52* (S4), doi:10.1086/658368

Zeder, M. A. (2008). Domestication and early agriculture in the Mediterranean Basin: origins, diffusion, and impact. *Proceedings of the National Academy of Sciences of the United States of America, 105* (33), 11597–11604. doi:10.1073/pnas.0801317105

Jakobsson, Mattias. Interview, December 2014.

McCartney, Carole. Field visit and interview, March 2014.

Museum of Cyprus, Nicosia. Visited in 2014.

Sotira Teppes. Visited in March 2014.

The Homes Built on the Graves of the Dead

It is possible to visit Çatalhöyük.

Hodder, I. (2006). *Çatalhöyük: the Leopard's tale: revealing the mysteries of Turkey's ancient 'town'*. London: Thames & Hudson.

Litt, T., Ohlwein, C., Neumann, F. H., Hense, A., & Stein, M. (2012). Holocene climate variability in the Levant from the Dead Sea pollen record. *Quaternary Science Reviews, 49,* 95–105. doi:10.1016/j.quascirev.2012.06.012

Paschou, P., Drineas, P., Yannaki, E., Razou, A., Kanaki, K., Tsetsos, F., et al. (2014). Maritime route of colonization of Europe. *Proceedings of the National Academy of Sciences of the United States of America, 111* (25), 9211–9216. doi:10.1073/pnas.1320811111

Özdoğan, M. (2011). Archaeological evidence on the westward expansion of farming communities from Eastern Anatolia to the Aegean and the Balkans. *Current Anthropology, 52* (S4), S415–S430. doi:10.1086/658895

Berggren, Å. Email, March 2015.

Hodder, Ian. Email, May 2014 and December 2014.

Litt, Thomas. Email, June 2014.

Özdoğan, Mehmet. Email, May 2014.

Clashes in Pilsen and Mainz

Bánffy, E. (2010). *Early Neolithic Settlement and Burials at Alsonyek-Bataszek*, Kraców.

Bollongino, R., Nehlich, O., Richards, M., Orschiedt, J., Thomas, M., Sell, C., et al. (2013). 2000 years of parallel societies in Stone Age Central Europe. *Science 342,* (6157), 479–481. doi:10.1126/science.1245049

Brandt, G., Haak, W., Adler, C., Roth, C., Szécsényi-Nagy, A., Karimnia, S., et al. (2013). Ancient DNA reveals key stages in the formation of central European mitochondrial genetic diversity. *Science*, *342* (6155), 257–261. doi:10.1126/science.1241844

Fernández, E., Pérez-Pérez, A., Gamba, C., Prats, E., Cuesta, P., Anfruns, J., et al. (2014). Ancient DNA analysis of 8000 B.C. Near Eastern farmers supports an Early Neolithic pioneer maritime colonization of mainland Europe through Cyprus and the Aegean Islands. *PLOS Genetics*, *10* (6), 1–16. doi:10.1371/journal.pgen.1004401

Alt, Kurt. Interview, October 2013.

Banffy, Eszter. Email, May 2014.

Brandt, Guido. Interview, October 2013.

Burger, Joachim. Interview, October 2013.

European Association for Archaeology. Annual meeting, September 2013.

Sowing and Sunrise

The State Museum of Prehistory (Landesmuseum für Vorgeschichte) in Halle is one of the best museums of prehistory in the whole of Europe, comparable only with Les-Eyzies-de-Tayac and the Danish National Museum in Copenhagen.

Unfortunately, the city of Halle is not equally agreeable. I recommend staying overnight in Berlin, or, even better, in Leipzig, which is only half an hour's journey by local train. From there, you can visit the Goseck solar observatory (and the Nebra Ark, described in 'The Nebra sky disc in Halle'). It is possible to reach Goseck, albeit rather circuitously, by local train and bus, but I hired a car to get there quickly and safely.

To experience the true magic of the solar observatory, you need to try and arrange to be there at sunrise or sunset on a clear day.

Zich, Bernhard. Interview, September 2013.

Archaeological Museum, Cracow. Visited in October 2013.

Goseck. Visited in November 2014.

State Museum of Prehistory (Landesmuseum für Vorgeschichte), Halle. Visited in September 2013 and November 2014.

Farmers Arrive in Skåne

The best place to study the arrival of agriculture in Scandinavia is the Danish National Museum in Copenhagen, the workplace of archaeologist Lasse Sørensen.

His new three-volume doctoral thesis complements the new DNA findings well, for readers who wish to study the subject in some depth.

Jennbert, K. (1984). *Den produktiva gåvan: tradition och innovation i Sydskandinavien för omkring 5300 år sedan*. Diss. Lund: University of Bonn.

Skoglund, P., Malmström, H., Omrak, A., Raghavan, M., Valdiosera, C., Günther, T., et al. (2014). Genomic diversity and admixture differs for Stone-Age Scandinavian foragers and farmers. *Science, 344* (6185), 747–750. doi:10.1126/science.1253448

Sørensen, L. & Karg, S. (2012). The expansion of agrarian societies towards the North – new evidence for agriculture during the Mesolithic/Neolithic transition in Southern Scandinavia. *Journal of Archaeological Science*, doi:10.1016/j.jas.2012.08.042

Sørensen, L. (2014). *From hunter to farmer in Northern Europe. Vol. 1–3: migration and adaptation during the Neolithic and Bronze Age*. Diss. Copenhagen: Oxford.

Andersson, Magnus. Interview, February 2014.

Broström, Anna. Interview, February 2014.

Jennbert, Kristina. Interview, February 2014.

Magnell, Ola. Interview, February 2014.

Ötzi the Iceman

I visited the South Tyrol Museum of Archaeology in Bolzano (Bozen) during a skiing holiday a few years ago. The extra train journey, which took a few hours, proved well worth the effort. Ötzi, the well-preserved ice mummy, along with his clothes and his gear, provides insights into the prehistoric past that it would be difficult to come by in any other way.

DeSalle, R. & Grimaldi, D. (1994). Very old DNA. *Current Opinion in Genetics & Development, 4*, 810–815. doi:10.1016/0959-437x(94) 90064-7 Ermini, L., Olivieri, C., Luciani, S., Marota, I., Rollo, F., Rizzi, E., et al. (2008). Complete mitochondrial genome sequence of the Tyrolean Iceman. *Current Biology, 18* (21), 1687–1693. doi:10.1016/j.cub.2008.09.028

Keller, A., Graefen, A., Ball, M., Matzas, M., Boisguerin, V., Maixner, F., et al. (2012). New insights into the Tyrolean Iceman's origin and phenotype as inferred by whole-genome sequencing. *Nature Communications, 3*, 698. doi:10.1038/ncomms1701

Müller, W., Fricke, H., Halliday, A., McCulloch, M., & Wartho, J. (2003). Origin and migration of the Alpine Iceman. *Science*, *302* (5646), 862–866.

Sikora, M., Carpenter, M. L., Moreno-Estrada, A., Henn, B. M., Underhill, P. A., Sánchez-Quinto, F., et al. (2014). Population genomic analysis of ancient and modern genomes yields new insights into the genetic ancestry of the Tyrolean Iceman and the genetic structure of Europe. *PLOS Genetics*, *10* (5), 1–12. doi:10.1371/journal.pgen.1004353

Spindler, K. (2000). *Mannen i isen*. Stockholm: Natur & Kultur.

Sjövold, Torstein. Email, March 2015.

South Tyrol Museum of Archaeology, Bolzano (Bozen). Visited in December 2009.

The Falbygden Area

The Falbygden Museum in Falköping is small, but pleasant and interesting. Exhibits include one of the world's oldest dogs and the mysterious Raspberry Girl. A replica of a stone burial place gives a good idea of how they were constructed.

The 'prehistoric village' at Ekehagen (Ekehagens forntidsby) brings history to life for both children and adults.

I would recommend that visitors to the Falbygden region fit in a visit to 'naturum Hornborgasjön', an information centre at Lake Hornborga, which provides information not only about cranes and other birdlife, but also about the first inhabitants of Sweden during the Palaeolithic period.

See Chapters 14 and 22. In addition, the following may be of interest:

Fu, Q., Pääbo, S., Rudan, P., & Krause, J. (2012). Complete mitochondrial genomes reveal neolithic expansion into Europe. *PLOS ONE*, 7 (3), doi:10.1371/journal.pone.0032473

Gamba, C., Jones, E. R., Teasdale, M. D., McLaughlin, R. L., Gonzalez-Fortes, G., Mattiangeli, V., et al. (2014). Genome flux and stasis in a five millennium transect of European prehistory. *Nature Communications*, 5 (10), 5257. doi:10.1038/ncomms6257

Isaksson, S. & Hallgren, F. (2012). Lipid residue analyses of Early Neolithic funnel-beaker pottery from Skogsmossen, eastern Central Sweden, and the earliest evidence of dairying in Sweden. *Journal of Archaeological Science*, *39*, 3600–3609. doi:10.1016/j.jas.2012.06.018

Malmström, H., Linderholm, A., Skoglund, P., Storå, J., Sjödin, P., Gilbert, M. P., et al. (2014). Ancient mitochondrial DNA from the northern fringe of the Neolithic farming expansion in Europe sheds light on the dispersion process. *Philosophical Transactions of the Royal Society of London. Series B, Biological Sciences, 370,* doi:10.1098/rstb.2013.0373

Salque, M., Bogucki, P. I., Pyzel, J., Sobkowiak-Tabaka, I., Grygiel, R., Szmyt, M., & Evershed, R. P. (2013). Earliest evidence for cheese making in the sixth millennium B.C. in Northern Europe. *Nature, 493* (7433), 522–525. doi:10.1038/nature11698

Wilde, S., Timpson, A., Karola, K., Kaiser, E., Kayser, M., Unterländer, M., et al. (2014). Direct evidence for positive selection of skin, hair, and eye pigmentation in Europeans during the last 5,000 years. *Proceedings of the National Academy of Sciences of the United States of America, 111* (13), 4832–4837. doi:10.1073/pnas.1316513111

Jakobsson, Mattias. Interview, November 2014.

Ekehagen prehistoric village (Ekehagens forntidsby). Visited in July 2014.

Museum of the Falbygden region (Falbygdens museum). Visited in July 2014.

Hunters' and Farmers' Genes

Axelsson, E., Ratnakumar, A., Arendt, M., Maqbool, K., Webster, M. T., Perloski, M., et al. (2013). The genomic signature of dog domestication reveals adaptation to a starch-rich diet. *Nature, 495* (7441), 360–364. doi:10.1038/nature11837

Bos, K. I., Harkins, K. M., Herbig, A., Coscolla, M., Weber, N., Comas, I., et al. (2014). Pre-Columbian mycobacterial genomes reveal seals as a source of New World human tuberculosis. *Nature, 514* (7523), 494–497. doi:10.1038/nature13591

Freedman, A. H., Gronau, I., Schweizer, R. M., Ortega-Del Vecchyo, D., Han, E., Silva, P. M., et al. (2014). Genome sequencing highlights the dynamic early history of dogs. *PLOS Genetics, 10* (1), 1–12. doi:10.1371/journal.pgen.1004016

Henry, A. G., Brooks, A. S., & Piperno, D. R. (2014). Plant foods and the dietary ecology of Neanderthals and early modern humans. *Journal of Human Evolution, 69* 44–54. doi:10.1016/j.jhevol.2013.12.014

Lazaridis, I., Patterson, N., Mittnik, A., Renaud, G., Mallick, S., Sudmant, P. H., et al. (2013). Ancient human genomes suggest three ancestral populations for present-day Europeans. *Nature*, *513*, 409–413. doi:10.1038/nature13673

Marlowe, F. W., Berbesque, J. C., Wood, B., Crittenden, A., Porter, C., & Mabulla, A. (2014). Honey, Hadza, hunter-gatherers, and human evolution. *Journal of Human Evolution*, 71, 119–128. doi:10.1016/j.jhevol.2014.03.006

Mathiesen, I. (2015). Eight thousand years of natural selection in Europe. *bioRxiv* preprint, first published online on 14 March 2015.

Perry, G. H., Dominy, N. J., Claw, K. G., Lee, A. S., Fiegler, H., Redon, R., et al. (2007). Diet and the evolution of human amylase gene copy number variation. *Nature Genetics*, *39* (10), 1256–1260. doi:10.1038/ng2123

Revedin, A., Aranguren, B., Becattini, R., Longo, L., Marconi, E., Lippi, M. M., et al. (2010). Thirty thousand-year-old evidence of plant food processing. *Proceedings of the National Academy of Sciences of the United States of America*, *44*, 18815–18819. doi:10.1073/pnas.1006993107.

Henry, Amanda. Telephone interview, November 2014

Lindblad-Toh, Kerstin. Interview, December 2014.

The First Stallion

Anthony, D. W. (2007). *The Horse, the Wheel, and Language: how Bronze-Age riders from the Eurasian steppes shaped the modern world.* Princeton, N.J.: Princeton University Press.

Ardvidsson, S. (2006) *Aryan idols: Indo-European mythology as ideaology and science.* Chicago: University of Chicago Press.

Bendrey, R. (2012). From wild horses to domestic horses: a European perspective. *World Archaeology*, *44* (1), 135–157. doi:10.1080/004 38243.2012.647571

Gamba, C., Jones, E. R., Teasdale, M. D., McLaughlin, R. L., Gonzalez-Fortes, G., Mattiangeli, V., et al. (2014). Genome flux and stasis in a five millennium transect of European prehistory. *Nature Communications*, *5*, 5257. doi:10.1038/ncomms6257

Gimbutas, M. (1989). *The Language of the Goddess: unearthing the hidden symbols of Western civilization.* London: Thames & Hudson.

Lindgren, G., Backström, N., Hellborg, L., Einarsson, A., Vilà, C., Ellegren, H., et al. (2004). Limited number of patrilines in horse domestication. *Nature Genetics*, *36* (4), 335–336. doi:10.1038/ng1326

Mallory, J. P. (1989). *In Search of the Indo-Europeans: language, archaeology and myth.* London: Thames & Hudson.

Mallory, J. P. & Adams, D. Q. (2006). *The Oxford Introduction to Proto-Indo-European and The Proto-Indo-European World*. Oxford: Oxford University Press.

Outram, A., Kasparov, A., Stear, N., Evershed, R., Bendrey, R., Olsen, S., et al. (2009). The earliest horse harnessing and milking. *Science*, *323* (5919), 1332–1335. doi:10.1126/science.1168594

Vilà, C., Leonard, J. A., Götherström, A., Marklund, S., Sandberg, K., Lidén, K., et al. (2001). Widespread origins of domestic horse lineages. *Science*, *291* (5503), 474–477.

Warmuth, V., Eriksson, A., Bower, M., Barker, G., Barrett, E., Hanks, B., et al. (2012). Reconstructing the origin and spread of horse domestication in the Eurasian steppe. *Proceedings of the National Academy of Sciences of the United States of America*, *109* (21), 8202–8206. doi:10.1073/pnas.1111122109

Anthony, David. Email, September 2014.

Eriksson, Anders. Email, September 2014.

Kveiborg, Jacob. Email, May 2015.

Mallory, James P. Interview, January 2015.

Vandkilde, Helle. Email, May 2015.

DNA Sequences Provide Links with the East

It was tricky writing the chapter about DNA and the arrival in western Europe of the herders from the steppe, as this is an area of cutting-edge research. The target kept on advancing rapidly throughout the time that I was writing the book.

The groundbreaking Boston study by Lazaridis et al. was only published on *Nature*'s site in September 2014. Fortunately, I was able to read a preliminary version several months in advance.

The most decisive study, which amounts to a revision of European history, is Haak et al., also from Boston. This was published in *Nature* in March 2015, by which time I had already submitted my manuscript. Still later, in June 2015, the rival team from Copenhagen and Gothenburg was able to confirm the scenario of the dispersal of the Indo-European languages through migration from the steppes in the east. Batini et al. (May 2015) shows, on the basis of Y chromosomes from men living today, that two thirds of the patrilineages in Europe can be traced back to these Bronze Age clans.

So a number of competing research teams that are world leaders in their field have now shown how the population of Europe changed at the dawn of the Bronze Age.

The State Museum of Prehistory (Landesmuseum für Vorgeschichte) in Halle an der Saale, Germany, is another excellent source of information about this period.

Both James P. Mallory's *In Search of Indo-European Languages* and David W. Anthony's *The Horse, the Wheel, and Language* are worth reading. However, even though many of Mallory's and Anthony's hypotheses have proven to be correct, their books are now slightly outdated in the light of the latest DNA findings.

Allentoft, M. E., Sikora, M., Sjögren, K., Rasmussen, S., Rasmussen, M., Stenderup, J., et al. (2015). Population genomics of Bronze Age Eurasia. *Nature*, *522* (7555), 167–172. doi:10.1038/nature14507

Batini, C., Hallast, P., Zadik, D., Delser, P. M., Benazzo, A., Ghirotto, S., et al. (2015). Large-scale recent expansion of European patrilineages shown by population resequencing. *Nature Communications*, *6*, 7152. doi:10.1038/ncomms8152

Chunxiang, L., Hongjie, L., Yinqiu, C., Chengzhi, X., Dawei, C., Wenying, L., et al. (2010). Evidence that a West-East admixed population lived in the Tarim Basin as early as the early Bronze Age. *BMC Biology*, *8*, 15. doi:10.1186/1741-7007-8-15

Haak, W., Brandt, G., de Jong, H. N., Meyer, C., Ganslmeier, R., Heyd, V., et al. (2008). Ancient DNA, Strontium isotopes, and osteological analyses shed light on social and kinship organization of the later Stone Age. *Proceedings of the National Academy of Sciences of the United States of America*, *47*, 18226–18231.

Haak, W., Lazaridis, I., Patterson, N., Rohland, N., Mallick, S., Llamas, B., et al. (2015). Massive migration from the steppe was a source for Indo-European languages in Europe. *Nature*, *522* (7555), 207–211. doi:10.1038/nature14317

Keyser, C., Bouakaze, C., Crubézy, E., Nikolaev, V. G., Montagnon, D., Reis, T., et al. (2009). Ancient DNA provides new insights into the history of south Siberian Kurgan people. *Human Genetics*, *126* (3), 395–410. doi:10.1007/s00439-009-0683-0

Lazaridis, I., Mallick, S., Nordenfelt, S., Li, H., Rohland, N., Economou, C., et al. (2014). Ancient human genomes suggest three ancestral populations for present-day Europeans. *Nature*, *513* (7518), 409–413. doi:10.1038/nature13673

Underhill, P., Poznik, G., Rootsi, S., Järve, M., Lin, A., Wang, J., et al. (2014). The phylogenetic and geographic structure of Y-chromosome haplogroup R1a. *European Journal of Human Genetics*, *23*, 124–131. doi:10.1038/ejhg.2014.50

Battleaxes

Allentoft, M. E., Sikora, M., Sjögren, K., Rasmussen, S., Rasmussen, M., Stenderup, J., et al. (2015). Population genomics of Bronze Age Eurasia. *Nature*, *522* (7555), 167–172. doi:10.1038/nature14507

Haak, W., Lazaridis, I., Patterson, N., Rohland, N., Mallick, S., Llamas, B., et al. (2015). Massive migration from the steppe was a source for Indo-European languages in Europe. *Nature*, *522* (7555), 207–211. doi:10.1038/nature14317

Larsson, Å. M. (2009). *Breaking and making bodies and pots: material and ritual practices in Sweden in the third millennium BC*. Diss. Uppsala: University of Uppsala.

Malmer, M. P. (2002). *The Neolithic of South Sweden: TRB, GRK, and STR*. Stockholm: Royal Swedish Academy of Letters, History and Antiquities.

Mathiesen, I. (2015) Eight thousand years of natural selection in Europe. *bioRxiv* preprint, first published online in 14 March 2015.

Brink, Kristian. Telephone interview, September 2014.

Kristansen, Kristian. Interview, August 2014.

Larsson, Lars. Interview, August 2013.

Larsson, Åsa M. Email, September 2014.

Sjögren, Karl-Göran. Interview, August 2014.

Bell Beakers, Celts and Stonehenge

I had the privilege of touring Stonehenge with Pat Shelley, a knowledgeable guide from Salisbury and Stonehenge Guided Tours.

Several times a year he leads special tours that start at sunrise and continue till lunchtime. You can actually go inside the stone circles, which visitors are not normally allowed to do. One exception is at midsummer, when the public are allowed right up to the monument – but then the area is far too crowded. Pat Shelley's special tours start in the early morning from nearby Salisbury. For practical reasons, it may be advisable to spend the night there; this also has the advantage of allowing some time to visit Salisbury Museum, where the Amesbury Archer is displayed.

It is best not to visit Stonehenge in the high season, if you have any choice in the matter. Spring, autumn and winter are better. If you visit in winter, don't make my mistake of not dressing warmly enough!

Anthony, D. W. (2007). *The Horse, the Wheel, and Language: how Bronze-Age riders from the Eurasian steppes shaped the modern world.* Princeton, N.J.: Princeton University Press.

Clark, P. (ed.) (2009). *Bronze Age Connections: cultural contact in prehistoric Europe.* Oxford: Oxbow Books.

Lee, E. J., Makarewicz, C., Renneberg, R., Harder, M., Krause-Kyora, B., Müller, S., et al. (2012). Emerging genetic patterns of the European Neolithic: perspectives from a late Neolithic Bell Beaker burial site in Germany. *American Journal of Physical Anthropology, 148* (4), 571–579. doi:10.1002/ajpa.22074

Mallory, J. P. (1989). *In Search of the Indo-Europeans: language, archaeology and myth.* London: Thames & Hudson.

Myres, N., Rootsi, S., Järve, M., Kutuev, I., Pshenichnov, A., Yunusbayev, B., et al. (2011). A major Y-chromosome haplogroup R1b Holocene era founder effect in Central and Western Europe. *European Journal of Human Genetics, 19* (1), 95–101. doi:10.1038/ejhg.2010.146

Parker Pearson, M. (2012). *Stonehenge: exploring the greatest Stone Age mystery.* London: Simon & Schuster.

Parker Pearson, M. (2015). *Stonehenge: making sense of a prehistoric mystery.* York: Council for British Archaeology.

Richards, J. C. (2007). *Stonehenge: the story so far.* Swindon: English Heritage.

Wessex Archaeology. (2014). The Amesbury Archer. Downloaded on 25 September 2014 from www.wessexarch.co.uk/book/export/html/5

Gaffney, Vincent. Interview, January 2015.

Shelley, Pat. Guide at Salisbury and Stonehenge Guided Tours.

Underhill, P. Email, September 2014

Stonehenge. Visited in January 2015.

The Nebra Sky Disc in Halle

The Nebra sky disc is the main attraction at the excellent State Museum of Prehistory (Landesmuseum fur Vorgeschichte) in Halle an der Saale (see previous chapters).

There is also a good exhibition at a purpose-built centre near the site of the find. The Nebra Ark (Arche Nebra) is just outside the town of Nebra. The nearest train station is at Wangen; the nearest bus stop is marked Großwangen.

Kristiansen, K. (1989). Prehistoric migrations – the case of the Single Grave and Corded Ware cultures. *Journal of Danish Archaeology, 8* (1), 221–225. doi:10.1080/0108464X.1989.10590029

Lappalainen,T., Hannelius, U., Salmela, E., von Döbeln, U., Lindgren, C. M., Huoponen, K., et al. (2009). Population structure in contemporary Sweden – a Y-chromosomal and mitochondrial DNA analysis. *Annals of Human Genetics, 73* (1), 61–73. doi:10.1111/j.1469-1809.2008.00487.x

Ling, J., Hjärthner-Holdar, E., Grandin, L., Billström, K., & Persson, P. (2013). Moving metals or indigenous mining? Provenancing Scandinavian Bronze Age artefacts by lead isotopes and trace elements. *Journal of Archaeological Science, 40,* 291–304. doi:10.1016/j.jas.2012.05.040

Ling, J., Stos-Gale, Z., Grandin, L., Billström, K., Hjärthner-Holdar, E., & Persson, P. (2014). Moving metals II: provenancing Scandinavian Bronze Age artefacts by lead isotope and elemental analyses. *Journal of Archaeological Science, 41,* 106–132. doi:10.1016/j.jas.2013.07.018

Pokutta, D. A. (2013). *Population dynamics, diet and migrations of the Unetice culture in Poland.* Diss. Gothenburg: University of Gothenburg.

Thornton, C. & Roberts, B. (2009). Introduction: The beginnings of metallurgy in global perspective. *Journal of World Prehistory, 22* (3), 181–184. doi:10.1007/s10963-009-9026-2

YFull.com. Read on 25 May 2015.

Zich, Bernd. Interview, September 2013.

Arche Nebra, visitors' centre. Visited in November 2014.

State Museum of Prehistory (Landesmuseum für Vorgeschichte), Halle. Visited in September 2013 and November 2014.

Vitlycke Museum, Tanum. Visited in August 2014.

The Rock Engravers

I strongly recommend a visit to the Vitlycke Museum of Rock Carvings (Hällristningsmuseet) at Vitlycke near Tanumshede, Bohuslän. Some of the most remarkable petroglyphs are to be found just a few minutes' walk from the museum. It's best to set aside a whole day, so as to have enough time to walk, cycle or drive around the area and look at several rock engravings. The museum hires out bicycles. It takes between two and three hours by coach and/or train to get there from Gothenburg. The final stretch is between one and five kilometres (⅔–3 miles) on foot, depending on your connection.

Bengtsson, L. (2004). Bilder vid vatten. *Licenciate* thesis. Gothenburg: Univ. Gothenburg.

Kristiansen, K. (1998). *Europe Before History.* Cambridge: Cambridge University Press.

Kristiansen, K. & Larsson, T. B. (2005). *The Rise of Bronze Age Society: travels, transmissions and transformations.* Cambridge: Cambridge University Press.

Ling, J. (2008). *Elevated rock art: towards a maritime understanding of Bronze Age rock art in northern Bohuslän, Sweden.* Diss. Gothenburg: University of Gothenburg.

McGovern, P. E., Hall, G. R., & Mirzoian, A. (2013). A biomolecular archaeological approach to "Nordic grog". *Danish Journal of Archaeology, 2* (2), 112–131.

Rosengren Pielberg, G., Golovko, A., Fitzsimmons, C., Lindblad-Toh, K., Andersson, L., Sundström, E., et al. (2008). A *cis*-acting regulatory mutation causes premature hair graying and susceptibility to melanoma in the horse. *Nature Genetics, 40* (8), 1004–1009. doi:10.1038/ng.185

Ling, Johan. Telephone interview, October 2014.

Iron and the Plague

Bos, K. I., Schuenemann, V. J., Golding, G. B., Burbano, H. A., Waglechner, N., Coombes, B. K., et al. (2011). A draft genome of *Yersinia pestis* from victims of the Black Death. *Nature, 478* (7370), 506–510. doi:10.1038/nature10549

Gräslund, B. (2007). Fimbulvintern, Ragnarök och klimatkrisen år 536–537 e. Kr. Saga och sed.

Haensch, S., Bianucci, R., Signoli, M., Rajerison, M., Schultz, M., Kacki, S., et al. (2010). Distinct clones of *Yersinia pestis* caused the Black Death. *PLOS Pathogens, 6* (10), e1001134. doi:10.1371/journal. ppat.1001134

Harbeck, M., Seifert, L., Hänsch, S., Wagner, D. M., Birdsell, D., Parise, K. L., et al. (2013). *Yersinia pestis* DNA from skeletal remains from the 6th century AD reveals insights into Justinianic Plague. *PLOS Pathogens, 9* (5), 1–8. doi:10.1371/journal. ppat.1003349

Lagerås, P. (2013). Agrara fluktuationer och befolkningsutveckling på sydsvenska höglandet tolkade utifrån röjningsrösen. *Fornvännen*, *108* (2013), 263–277.

Larsen, L. B., Vinther, B. M., Briffa, K. R., Melvin, T. M., Clausen, H. B., Jones, P. D., et al. (2008). New ice core evidence for a volcanic cause of the A.D. 536 dust veil. *Geophysical Research Letters*, *35* (4), doi:10.1029/2007GL032450

Manco, J. (2013). *Ancestral Journeys: the peopling of Europe from the first ventures to the Vikings*. London: Thames & Hudson.

Myrdal, J. (2003). *Digerdöden, pestvågor och ödeläggelse: ett perspektiv på senmedeltidens Sverige*. Stockholm: Sällsk. Runica et Mediævalia.

Wagner, D. M., Klunk, J., Harbeck, M., Devault, A., Waglechner, N., Sahl, J. W., et al. (2014). *Yersinia pestis* and the plague of Justinian 541–543 AD: a genomic analysis. *The Lancet Infectious Diseases*, *14* (4), 319–326. doi:10.1016/S1473-3099(13)70323-2

Am I a Viking?

The Swedish History Museum in Stockholm houses a major exhibition on the Vikings. Birka, a former Viking settlement on the island of Björkö in Lake Mälaren, is a very popular tourist destination.

The Danish National Museum in Copenhagen is also strong on this period. The Viking Ship Museum in Oslo (Vikingskipshuset) and its counterpart in Roskilde, Denmark (Vikingeskibsmuseet) are other places of pilgrimage for Viking enthusiasts.

The books of Richard Hall and Jean Manco portray the period from a British perspective. I strongly recommend the new book by the Yale professor Anders Winroth, who was kind enough to check the facts in the Viking chapter of this book.

Ebenesersdóttir, S., Sigurdsson, A., Stefánsson, K., Helgason, A., Sánchez-Quinto, F., & Lalueza-Fox, C. (2011). A new subclade of mtDNA haplogroup c1 found in Icelanders: evidence of pre-columbian contact? *American Journal of Physical Anthropology*, *144* (1), 92–99. doi:10.1002/ajpa.21419

Goodacre, S., Helgason, A., Nicholson, J., Southam, L., Ferguson, L., Hickey, E., & et al. (2005). Genetic evidence for a family-based Scandinavian settlement of Shetland and Orkney during the Viking periods. *Heredity*, *95* (2), 129–135. doi:10.1038/sj.hdy.6800661

Hall, R. (2007). *Exploring the World of the Vikings*. London: Thames & Hudson.

Hall, R. A. (2012). *Exploring the World of the Vikings* (1st paperback edition). London: Thames & Hudson.

Haywood, J. (1995). *The Penguin Historical Atlas of the Vikings*. London: Penguin.

Helgason, A., Sigurðardóttir, S., Nicholson, J., Sykes, B., Hill, E. W., Bradley, D. G., et al. (2000). Estimating Scandinavian and Gaelic ancestry in the male settlers of Iceland. *The American Journal of Human Genetics*, 67, 697–717. doi:10.1086/303046

Helgason, A., Hickey, E., Goodacre, S., Bosnes, V., Stefánsson, K., Ward, R., & Sykes, B. (2001). mtDNA and the islands of the North Atlantic: estimating the proportions of Norse and Gaelic ancestry. *The American Journal of Human Genetics*, 68, 723–737. doi:10.1086/318785

Krzewińska, M., Bjørnstad, G., Skoglund, P., Olason, P. I., Bill, J., Götherström, A., & Hagelberg, E. (2015). Mitochondrial DNA variation in the Viking age population of Norway. *Philosophical Transactions of the Royal Society of London. Series B, Biological Sciences*, 370 (1660), doi:10.1098/rstb.2013.0384

Leslie, S., Winney, B., Boumertit, A., Day, T., Hutnik, K., Royrvik, E., et al. (2015). The fine-scale genetic structure of the British population. *Nature*, 519 (7543), 309–314. doi:10.1038/nature14230

Manco, J. (2013). *Ancestral Journeys: the peopling of Europe from the first ventures to the Vikings*. London: Thames & Hudson.

Naumann, E., Krzewinska, M., Gotherstrom, A., & Eriksson, G. (2014). Slaves as burial gifts in Viking Age Norway? Evidence from stable isotope and ancient DNA analyses. *Journal of Archaeological Science*, 41, 533–540.

Winroth, A. (2014). *The Age of the Vikings*. Princeton: Princeton University Press.

Bojs, Anders. Conversations and emails, 2013–2015.

Salomonsson, Klas. Conversations and emails, 2013–2014.

Danish National Museum, Copenhagen. Visited in February 2014 and January 2015.

Swedish History Museum, Stockholm. Visited several times in 2014.

Urshult. Visited in the summer of 2013.

The Mothers

Karlberg, G. (1976[1908]). *Anteckningar om Glafva socken i Värmland* (new edition). Karlstad: Föreningen för Värmlandslitteratur.

Turesson, G. (1976). *Visor och skaldeminnen*. Stockholm: LT.

Andersson, Dagmar. Conversation, November 2014.

Juås, Birgitta. Conversation, July 2014.

Lindgren, K. Conversation, July 2012.

Olausson, Peter. Interview, July 2014.

Documents from Glava outdoor museum (*hembygdsgård*).

Further information has come to light since this book was written. If you think you could be related to me in the maternal line, please contact me via my homepage at karinbojs.se for more details.

The Legacy of Hitler and Stalin

Abecasis, G., Auton, A., Brooks, L. D., DePristo, M.A., et al. (2012). An integrated map of genetic variation from 1,092 human genomes. *Nature*, *491* (7422), 56–65. doi:10.1038/nature11632

Pringle, P. (2008). *The Murder of Nikolai Vavilov: the story of Stalin's persecution of one of the great scientists of the twentieth century.* New York: Simon & Schuster.

Jakobsson, Mattias. Interview, November 2014.

Pääbo, Svante. Interview, December 2014.

Vavilov Institute, St Petersburg. Visited in September 2010.

The Tree and the Spring

This chapter is a summary of the book's overall findings, together with my personal reflections. References to scientific and academic works are given for previous chapters. The legend of Saint Botvid – like all such legends – should be taken with a large pinch of salt. However, he is mentioned in a number of documents, including the Vallentuna Calendar or *Kalendarium Vallentunense* from the 1190s. More information about the church building (in Swedish) can be found on Botkyrka parish's homepage (www.visitbotkyrka.se).

Questions and Answers about DNA

If you are interested in a book containing more technical information about the use of DNA in researching family history, I recommend you read Magnus Bäckmark's *Genvägar*. It is important to get hold of the latest edition, as this kind of information dates quickly.

In 2015, the Swedish Association of Family History Researchers (Sveriges släktforskarförbund) also published a basic instruction manual by the expert Peter Sjölund, *Släktforska med DNA*.

Bäckmark, M. (2013). *Genvägar: praktisk handledning till DNA-jämförelse i släktforskning* (2nd, augmented edition). Åkersberga: Gröna stubben.

The International Society of Genetic Genealogy has plenty of useful information on its website: www.isogg.org

Acknowledgements

A very large number of people have played a part in bringing this book to fruition. As regards fact-checking alone, 33 people – all of them experts in their respective fields – have invested time and effort in combing through the text. Should any errors have crept in despite their efforts, the fault is mine, not theirs.

To list everyone who has been involved would take a disproportionate amount of space. The 70-odd researchers who kindly agreed to be interviewed and to reply to the floods of questions I sent them are listed in the references that precede these pages.

Nonetheless, there are a few individuals I would like to single out for special thanks. I am particularly grateful to Svante Pääbo, Pontus Skoglund, Mattias Jakobsson and Carole McCartney.

Among the genealogy researchers I have consulted, Håkan Skogsjö, Peter Sjölund and Mats Ahlgren in particular have gone out of their way to help me.

I would like to thank my uncle, Anders Bojs, for agreeing to take a DNA test.

My friends Christina Wilén, Daniel Olsson and Per Snaprud, who belong to my writers' circle, have taught me a great deal about the craft of storytelling, and my friend Anna Svensson has drilled me in how to cite references correctly.

I have benefited from the professional fact-checking and editorial expertise of Maria Gunther and Anna Bodin, my colleagues at *Dagens Nyheter*, while Stefan Rothmaier has done wonders with the book's illustrations.

A number of my journeys were financed by a grant from the Swedish Journalists' Association, *Svenska Journalistförbundet*.

The publishing staff at Albert Bonniers have had to put up with my uncompromising standards and exacting ambitions. I would like to express my appreciation of Martin Kaunitz's resolute interventions and Elisabeth Watson-Straarup's patience and language skills, which enabled her to gather illustrations from all over Europe.

And finally, I would like to thank my English translator, Fiona Graham, and the staff at Bloomsbury. It has been a pleasure to work with such professional people.

Index

Aboriginals 74, 82
Abri Pataud, France 63–64, 231, 232
acupuncture 210
Adam 23
agriculture 139–41, 230–31
    Anatolia, Turkey 185–87
    arrival in Sweden 203–205
    climate change 160–62, 181–82
    Cyprus 147–51, 168–71, 173–78
    expansion theory 205–206
    Galilee 158–60
    Hungary 187–88
    soil types 200–202
    Syria 145–46
Ahlström, Torbjörn 104
Ahrensberg culture 98–100
alcohol 167–68, 242, 250, 298–99
Alt, Kurt 192–93
amber 292–93
Amesbury Archer 275–76, 277,
    279, 289
Anatolia, Turkey 185–87
ancestors 10–11
Andersson, Dan 327, 328
Andersson, Leif 296
Anglo-Saxons 312
Anning, Mary 257
Anthony, David 223, 242–45, 254,
    262–64, 299–300
Aronsson, Kjell-Åke 126–29, 133
arrows 52, 92, 98, 106, 119, 135, 275
art 33, 42–43, 55–56, 75, 87
astronomy 197–99, 277–80
    Nebra sky disc 282–83, 284–86,
    287, 289
atlatls 65
Auel, Jean M. 245
    The Clan of the Cave Bear 28–29,
    54–55
Aurignacian culture 38–40, 50–51,
    56, 58–60, 63, 65, 69, 75

Baillie, Michael 302–303
Bánffy, Eszter 189–90, 265

Bar-Yosef, Ofer 158, 162–63
Barum woman 110–12, 118
Basques 178–80
Battleaxe culture 267–72
beads 50
Beckman, Lars 125
beer 164, 167–68, 204
Bell Beaker culture 266, 275–77,
    279–80, 281, 291, 300
Berbers 68
Beringia 86, 106
Bjerck, Hein Bjartmann 98–99
Black Death 303–306
Bleek, Wilhelm 76
boats 99–100, 101–102, 107, 263
    long-distance craft 290, 293, 309
bog people 227–29
Bohuslän, Sweden 98–101, 103–105,
    110–11, 203, 230
Bojs, Anita 9–10, 336–38
Bojs, Eric 10, 12, 313–15, 316
Bojs, Göran 314, 317
Bojs, Hilda 10, 12, 142, 146, 150,
    177, 187, 214, 215, 318–20
Bonn-Oberkassel, Germany 78–79,
    83, 85–87, 112, 161
Botai culture 239–40, 242
Brandt, Guido 191–95, 155, 265
Brno, Czech Republic 57–58
Bromme culture 92
bronze 288–89, 292–93, 301
    bronze axes 287–88
Burenhult, Göran 141
Burger, Joachim 193–95, 219
burials 32–33, 52
    Battleaxe culture 270
    Bell Beaker culture 281
    Bonn-Oberkassel, Germany 78–79,
    83, 85–87, 112, 161
    cat burials 154–55
    Corded Ware culture 256–57, 262,
    269, 281
    horse burials 242, 243
    lake burials 114–16

passage graves 214–15, 217, 223–27, 243, 261
Single Grave culture 271–72
Stonehenge, UK 275–76

canoes 98–99, 103, 119, 121–22, 148
Cardium culture 180
Çatalhöyük, Turkey 182–85, 221, 247
cats 153–56
    domestication 156–57
cave painting 37, 42–43, 69–70, 74–75
    Cap Blanc, France 70
    Font-de-Gaume, France 70–72, 74
    Les Combarelles, France 72–73
    Rouffignac, France 73–74
    shamanism 75–77
Celtic languages 277, 301
ceramics 60, 135–37, 145, 173–76, 180, 181, 222, 268–69, 275, 281
Châtelperronian culture 51–53
cheese 222
chewing gum 101
circular buildings 149–52
climate change 48–49, 54, 78, 82, 120–22
    agriculture 160–62, 181–82
    end of the Ice Age 86–87, 94–100
    Fimbulwinter 302–306
clothes 66
Clottes, Jean 75–77
Coles, Bryony 89
Comb Ware 138
Conard, Nicholas 41–43
Cook, Jill 41, 246–47
cooking pots 135–37
copper 211, 267–68, 288–89
Corded Ware culture 255–57, 260–62, 264, 266, 269, 271, 272–73, 275, 281, 291, 300
creativity 43–44, 337
Cro-Magnons 62–65, 232
    Cro-Magnon 1 64
cup marks 295
Cyprus 147–51, 168–71, 173–78

Dagens Nyheter 24, 84, 100, 164, 216
Darwin, Charles 257
    On the Origin of Species 62, 131

Dawkins, Richard The Selfish Gene 338
deCODE 33–36, 67–68, 122–23
Denisovans 26, 32
Dhra', Jordan 151–52, 155
Diamond, Jared 230
diseases 231
DNA 11–12
    ancient DNA 19–24, 45, 254
    ancient Siberians 106
    family history 345–47
    mitochondrial DNA 21–22, 35–37, 122–23, 128–29, 141–42, 146, 191–92, 193, 194–95, 254–55, 318, 322
    mutations 35–36, 44–45, 128, 259, 291, 318, 322–23
    Neanderthals 54
    nuclear DNA 24–25, 215, 217–218
    origins of Indo-European languages 263–66
    Ötzi the Iceman 211–13
    plant DNA 164–65
    Y chromosomes 23, 128, 212, 254, 255, 264, 273, 276–77, 280, 291, 310–13
Doggerland 86–91, 92–93
    Bromme culture 92
    flooding 120–22
    Hamburg culture 91–92
dogs 37, 61, 103, 156, 220
    Bonn-Oberkassel, Germany 78–79, 83, 85–87, 112, 161
    canine DNA 233–34
    dog meat 83–85
    hunting 82–83, 119
    wolves 79–82
Dolní Věstonice, Czech Republic 58–61, 79, 135, 232, 247

Ekehagen, Sweden 220–22
epigenetics 339
equinoxes 197, 284
Eriksdotter, Katarina 320–22
Erlandsson, Gudmund 333–35
Erlandsson, Märta 331, 332, 334, 335–36
Ertebølle culture 134–38

Eve 21–23, 35
eye colour 112, 117, 218, 266

family history 10, 12, 345–47
Family Tree DNA (FTDNA) 122–23,
128, 129, 258
fertility rites 246–47, 295–96
figurative art 42–43, 75
figurines 40–42, 135, 184, 242
Fimbulwinter 302–306
Finlayson, Bill 151–53
fish-hooks 106
flints 65, 206
flutes 39–40, 75
folk culture 248–49
Forest Finns 329–32
forests 118–20
FOXP2 gene 28, 32
Funnel Beaker culture 203–204,
206, 217, 224–27, 243, 261–62

Gaffney, Vincent 87–90, 278
Galilee 15, 17, 25, 31, 158–60
genetics 131, 189–90, 338–39
disputed research 340–45, 347
Gimbutas, Marija 245–48, 250–53,
254, 264, 270
Globular Amphora culture 262
Göbekli Tepe, Turkey 165–68,
204, 223
goddess theories 245–47, 251
Gökhem 4 140, 142, 214–15
gold 275, 276, 281–83, 289, 293
Götherström, Anders 140, 216
Gotland, Sweden 107, 116, 140, 203,
215, 217–19, 224
Gottfriedz, Berta 9–10, 141,
326–29, 336
granaries 151–53
Gravettian culture 58–59, 63,
64, 65
Great Migrations (Völkerwanderung)
301, 312
Gronenborn, Detlef 217
Gudmundsdotter, Karin 330–31,
333, 336
Gullbrandsdotter, Kajsa 327–29, 332
Günther, Hans F. 132–33, 252
Gyllensten, Ulf 126, 128

Haak, Wolfgang 142, 187, 265
hair colour 112, 117, 218
Hallgren, Fredrik 115, 265
Hamburg culture 91–92, 99
haplogroups 59
C1 310
G 212
H 142, 146, 193, 215, 320
H1g1 187–88
HV 146, 190
J 190
K 146, 190, 193, 211–12
N 146
N1a 190
N3 313
R1 255, 257, 259, 264, 273,
276–77, 280, 291, 300, 313,
317, 343
T2 190
U2 47, 59, 116, 142
U4 107, 109, 141–42, 193
U5 35–36, 47, 58, 67–68, 107,
123, 124, 141, 142, 190, 193
U5b1 68, 78–79, 85, 112
U5b1b 123, 124–26, 128, 331–32
U8 58, 193
V 125–26
W 190
X 190
Z 126, 128
Helena 320, 322
Henry, Amanda 53, 232–33
Herodotus 154
Heun, Manfred 164–65, 167
Higham, Tom 52
Hjortspring boat, Denmark 290, 294,
309
Hodder, Ian 184
Hohle Fels, Germany 38–42, 49,
184, 246–47
Holocaust 130, 132, 133, 248–49, 253
Homo floriensis 32
heidelbergensis 56
sapiens 32, 231
horses 239–42, 262–63, 266,
299–300
horse burials 242, 243
riding 242–43, 244
sacrifices 295–96

Hublin, Jean-Jacques   30–32, 52–53, 54, 56
Hummerviken, Norway   107, 109
Hungary   187–88
hunting   52–54, 60, 134–35
    hunting of marine mammals   97–98, 102, 107–108
    hunting with dogs   82–83, 119
Huseby Klev, Sweden   100–103, 105

Ibn Fadlan, Ahmad   323–24
Ice Age   38, 43, 48, 54, 58, 59, 61
    Cro-Magnons   62
    Doggerland   86–91
    dogs   80–83, 85
    end of the Ice Age   86–87, 94–100, 162
immune system   26–27
indigenous Americans   106, 158, 310
Indo-European languages   244–45, 248–52, 254, 263, 271, 292, 302
    DNA studies   263–66
Inuit   97–98, 99, 103
iron   127–28, 301–302

Jakobsson, Mattias   140, 216, 342, 344–45
Jakucs, János   190
Jeansdotter, Annika   335, 336
jewellery   50, 60, 111, 163, 270
Johansson, Per   316
Journal of Indo-European Studies   251, 252
Justinian plague   304–305

Kadrow, Sławomir   260–63
Karg, Sabine   204–205
Karlsson, Jon Love   44
Khirokitia, Cyprus   149–51
Khvalynsk culture   243
Kongemose culture   134
Kostenki-14   47–51, 59
Krause, Johannes   64, 265
Kristiansen, Kristian   270–72, 292
Kuijit, Ian   151–53

lactose   212, 221–22
Lagerås, Per   305, 306
Lagerlöf, Selma   333–34

lamps   137
language   28, 32
languages   127, 178–80, 244–45, 301–2
    Celtic languages   277, 301
Larsson, Åsa   268–70, 271
Larsson, Lars   82, 268, 271
Laugerie Haute, France   68–70
Lewis-Williams, David   75–77
Lindblad-Toh, Kerstin   233–34
Lindgren, K.-G   330–31
Lindisfarne, UK   308
Linear Band Ware   138
Linear Pottery (LBK) culture   193, 196–99, 201, 206, 212
Linear Ware culture   188
Ling, Johan   287–88, 293–94
Linnaeus   130, 223
Lion Man   40–41, 247
Lundborg, Herman   132–33, 344–45
Lysenko, Trofim   341–42, 327

Magdalenian culture   69, 71, 73–74, 78–79, 137
Mal'ta, Siberia   106, 218
Mallory, James   250–53, 254, 264, 302
Malmer, Mats   267–68, 273
Månsson, Måns   315–16, 317
McCartney, Carole   150–51, 173–74, 246
McGovern, Patrick   299
mead   250, 256–57, 260, 299
megaliths   223–24
    Stonehenge, UK   275, 277–80
Meller, Harald   282–83
Mendel, Gregor   57, 131
metalworking   127–28, 275, 289
mice   28, 151, 153, 155, 157, 163–64, 170–71, 184
migrationism   189–90, 224, 248
Milankovitch cycles   48, 86
milk   212, 221–22
miscegenation   130–31
mitochondrial DNA   21–22, 35–37, 122–23, 128–29, 141–42, 146, 191–92, 193, 194–95, 254–55, 318, 322
Moberg, Vilhelm   314
Mossberg, Bo   164

Motala, Sweden   114–16, 118, 218, 265
multi-regional hypothesis   19, 23
mummies   20–21, 24
Musée National de Préhistoire, Les
   Eyzies-de-Tayac, France   64–65
Museum of Prehistory, Halle,
   Germany   196
music   33, 37, 43, 55
   ancient flutes   39–40, 75
mutations   35–36, 44–45, 128, 259,
   291, 318, 322–23

Native Americans   41, 75–76, 109
Natufians   161–62
Nature   21, 106, 254, 263–66, 272, 342
Nature Communications   215
Neanderthals   18–19, 20, 24
   Châtelperronian culture   51–53
   extinction   32, 49–50, 52, 53–56
   fictional accounts   28–30
   Hohle Fels cave   38
   inbreeding   54
   interbreeding with modern
      humans   25–28, 30, 31–32, 47,
      55, 158–59, 261
Nebra horde   281–82, 283
   Nebra sky disc   282–83, 284–86,
      287, 289
needles   37, 49, 65–66
Nilsson, Peter   316
Nilsson, Ruben   327
non-figurative art   42
Nordqvist, Bengt   101, 103
nuclear DNA   24–25, 215, 217–218

obsidian   163, 166, 185
Olausson, Peter   332–35
Öresund Strait   91–93, 99, 122
Österöd woman   103–105, 230
Ötzi the Iceman   208–11, 215,
   221, 289
   DNA analysis   211–13
oxen   200–201, 207, 222, 224–25,
   243–44, 262

Pääbo, Svante   19–24, 24–28, 45, 55,
   219, 232, 342, 343–44
Palaeolithic diets   231–33
   Paleo diet   231, 233

Palme, Olof   313
Parham, Peter   26
Patterson, Nick   265
PCR (polymerase chain reaction)   23
Pearson, Roger   252
Persson, Johan   316
pigments   52, 71–73, 78, 111, 149,
   183, 196, 218
plague   303–306
Plassard, Frédéric   74–75
ploughs   200–202, 207, 224–25, 249
Pokutta, Dalia   285–86
Potter, Beatrix   257
pottery   135–38, 180
   Battleaxe culture   268–69, 270
   Corded Ware culture   269
   Cyprus   174–75
Procopius   304
psychosis   43–44, 45–46

racial biology   130–33, 344
radiocarbon dating   52, 55, 64, 105,
   139–40, 190, 204, 219
Ragnar   259–60, 273–74,
   291–92, 317
Raspberry Girl   227–29, 230
Reich, David   265, 342
reindeer   97–100
   domestication   126–27
rock carvings   293–97
Rurik   312–13

Samara culture   242–43
Sami   41, 68, 124–33, 332, 344–45
   languages   302–303
Samuelsson, Klas   315
San people   41, 76
Savolainen, Peter   83–85
schizophrenia   43–45
Science   140, 164, 192, 193, 194, 195,
   215, 216, 217, 224, 318, 342
shamanism   41, 75–77
   Sami   127, 129–30
Siberian nomads   41, 75
Single Grave culture   271–72
Sjögren, Karl-Göran   224–27, 268
Sjölund, Peter   250–51, 291, 313, 317
skin colour   112–14, 116–17, 218,
   228–29, 266

Skoglund, Pontus 140, 216
Skogsjö, Håkan 322, 329, 330
slave trading 293, 300, 307, 309,
   311, 323
solstices 197–99, 277–80
Solutrean culture 64–65, 68–71, 261
Sørensen, Lasse 140, 204–205
spears 52–53, 65, 87, 92, 119, 290
   spear-throwers 65
Stalin, Joseph 57, 253, 338, 341
Starčevo culture 188, 190
starchy foods 53, 159–60, 231–35
statuettes 40–42, 75, 298
Stefánsson, Kári 33–34
steppe cultures 239–45, 249–50, 252,
   254, 262–66, 270, 272–73, 300
Stonehenge, UK 275, 277–80, 289
   Durrington Walls 278–79
   Heel Stone 279
   Woodhenge 278
Stoneking, Mark 66
Stora Förvar, Sweden 107–109
Storå, Jan 140, 216
Storegga Slide 121–22
storytelling 43
Street, Martin 83–84, 119
Svensdotter, Annika 329–30
Svensdotter, Karin 329, 330
Svoboda, Jiří 59–61, 67
Swiderian culture 98
Sykes, Bryan 35–36
symmetry 55–56
Syria 145–46
Szécsényi-Nagy, Anna 265

Tambets, Kristiina 125
tin 288–89, 293
tools 50, 51, 55, 65, 87, 135
trade 287, 289–92, 300
Trundholm sun chariot,
   Denmark 297–98
Turesson, Gunnar 328, 329, 332
Turesson, Karolina 10, 327, 328, 329

Uluburun boat, Turkey 293
Uluzzian culture 51
Underhill, Peter 257, 258, 276
Unetice culture 281, 285–86
Ursula 35–37, 59
   haplogroups 67

Varna culture 289
Vavilov, Nikolai 340–42
Venus of Dolní Věstonice 60, 61
Venus of Hohle Fels 41–42, 184, 246
Vigilant, Linda 22
Vikings 306, 307–13, 323–25
   indigenous Americans 310
Vitlycke rock carvings,
   Sweden 293–97
volcanic eruptions 48–49, 51, 303

Wade, Nicholas 342
Wayne, Robert 81
wheat 164–65
wheels 200–202, 207, 224–25,
   243–44, 249, 262
Wilcox, George 163
Willerslev, Eske 342
Williams-Beuren syndrome 81
Wilson, Allan 21–22
wine 148, 154, 167, 299
Winroth, Anders 309
wolves 79–82
woollens 299–300

Y chromosomes 23, 128, 212, 254,
   255, 264, 273, 276–77, 280, 291,
   310–13
   Big Y 258–59
   YFull 291
Yamnaya culture 243–44, 264–66,
   272–73, 276
Younger Dryas 92–93, 162–62

Zich, Bernd 196–97, 200–201,
   206–207, 224, 261, 283–84